Being Successful As An Engineer

W. H. Roadstrum

Professor of Electrical Engineering and Management
Worcester Polytechnic Institute

Formerly Systems Engineer, Manager of Missile
Systems Engineering, Manager, Information Storage
and Retrieval, and Manager, Surveillance and De-
tection Engineering, General Electric Company;
Engineer, U.S. Bureau of Mines

Registered Professional Engineer—State of New York

Engineering Press, Inc.   San Jose, California 95103

Library of Congress Cataloging in Publication Data

Roadstrum, William Henry, 1915-
    Being successful as an engineer.

    Bibliography: p.
    Includes index.
      1.   Engineering — Vocational guidance.
I.   Title.
TA157.R59   620'.0023   77-27435
ISBN 0-910554-24-2

Printed in the United States of America

ENGINEERING PRESS, INC.   **P.O. Box 5**   **San Jose, California 95103**

Preface

This book is intended primarily as a tool for the young engineer, who will be able to move quickly and effectively from college into industry with the aid of the ideas discussed and the practical paths toward expertness in the engineering art that are outlined. The experience-based suggestions and helpful warnings will allow the engineer to take the first steps into project leadership and group management more confidently.

It is essential to understand that engineering practice and the "engineering" of engineering schools differ greatly. Practice emphasizes application; education properly emphasizes technology. Practice organizes itself around creative, practical projects. Schooling is organized around predigested, theoretical courses. Engineering schooling almost inevitably biases its students toward theory and away from the practical — toward piecemeal analysis and away from project synthesis. Many, if not most, early career difficulties stem from a misunderstanding of these points. The engineer who clings to school attitudes slows his or her career and invites obsolescence.

It is the intent of this book that with its help a man or woman should gain more and gain faster from work experiences. The key to engineering success, pointed out chapter by chapter, is to relate each detail of practice to the fundamental nature and purpose of engineering itself. The reader is required to take a project point of view.

The book should be especially valuable to engineers establishing themselves in the years immediately following graduation. An initial reading will go far to dispel common illusions about practice. Later review may also be a helpful part of the periodic self-renewal efforts that every professional needs. The practical division of subject matter, detailed table of contents, and complete index will also encourage use of this book for reference on specific professional problems.

Grateful acknowledgment is made to the following individuals, organizations, and publishers for permission to quote: American Society for Engineering Education; Dr. Vannevar Bush; Engineers' Council for Professional Development; Francis K. McCune; McGraw-Hill Book Company; and Prentice-Hall. I am also grateful to John Wiley & Sons for early release of the material from my book, *Excellence in Engineering,* of which this book approximates a second edition. My chief indebtedness is to associates and superiors over many years whose ideas and examples have been consciously or unconsciously absorbed into my own thinking and practice. There are far too many to mention individually. But I thank them all.

W. H. Roadstrum

Worcester, Massachusetts

Contents

Chapter One

What Engineering Is

Attaining excellence in engineering practice is fairly simple. The engineer keeps clearly in view a broad mission or purpose and relates this to each of the details of everyday work as they arise. A person who is able to do this most of the time will be successful. We will want to examine carefully some of the seemingly complicated and subtle details. These fall into place with surprising ease, however, when related directly to the overall goal.

The reader's first task in this book, then, is to gain a clear understanding of what will be his or her own purpose as an engineer. This chapter lays a firm foundation for understanding engineering itself and the general nature of engineering practice. Chapter 2 encourages the engineer to look closely at the profession from a career and personal development standpoint, and chapters 15 and 17 discuss these personal aspects in more detail.

Beginning in chapter 3, the book takes up — chapter by chapter — the important elements of practice encountered in daily work and relates them to the broad picture presented in this chapter. The reader will see at once the project nature of practical engineering work. Relationships to other members of the project team and how to make them effective will become apparent as an understanding develops of what project control must be. Report-making, problem-solving, laboratory skills, studies and proposals, and systems work all fall neatly into place. The basic ideas of research and development, design, production, and marketing are seen to grow directly out of the nature of engineering presented in this chapter. The final chapters of the book include a look at creativity itself and the work of the engineering manager.

It is recommended that the engineer first entering practice read thoughtfully through the whole book. After this, as problems arise, reference may be made to specific chapters to help resolve them. As an engineer's responsibilities increase, chapters 14, 15, and 19 will become particularly meaningful.

Engineering Is a Need-Filling Activity

It is almost impossible for an engineer to take too big a look at anything. We will see throughout this book that the problem lies in exactly the opposite direction. Each of us is tempted to take too small and narrow an approach to both technical and everyday problems. During an engineer's career, the ability to take a long-perspective, overall look increases, especially when management responsibilities and responsibility to society grow.

Our history is a record of the struggle to satisfy several basic human needs. We usually classify them as food, clothing, and shelter. Today these three needs are the basic concern of most of this planet's inhabitants. Everyone knows what they mean. But there is a fourth and lesser-known need just as vital as the traditional three. It has always existed for everyone, though it may be disguised in various ways. I call this fourth need *fulfillment*. Some label it a *social* or *self-realization need.** Basically, it is the need for an individual to fulfill a role. Every person is impelled to live usefully, abundantly, successfully, worthily, purposefully, boldly, and beautifully.

As an engineer, you are going to be involved in providing for some basic human needs in various ways. Directly or indirectly, your work will help people to get more and better food, clothing, and shelter. But be careful that your efforts in this direction do not subtract from the satisfaction of the fourth, less tangible need. Engineer Othmar Amman, designer of the Verazanno Narrows Bridge in New York, said in an interview, "It is a crime to build an ugly bridge."

To examine this from another standpoint, suppose you are an industrial engineer improving the manufacturing process in your factory. Perhaps by some novel combination of machines and flow patterns you can significantly improve the *mechanical* efficiency of getting the work done. But if these "improvements" lead to a new organization of processes that prevents workers from seeing the results of their own efforts (depriving them of the satisfaction of accomplishment), theoretical gain in efficiency could be lost in poorer employee performance.

Engineering is different from other professions and occupations in that it *uses technical knowledge to meet human needs.* A merchant helps meet human needs through applying principles and techniques of retail business management. A physician attempts to meet human needs with principles from biology and psychology; and so on, for the minister, the laborer, the tailor, the lawyer. (We will take a look at the difference between professional and other work in the next chapter.) The engineer's work is not just technology (or technical knowledge)! It is a doing — an *application* of

---

* See, for example, Herbert G. Hicks and C. Ray Gullett, *The Management of Organizations,* 3rd ed., McGraw-Hill, New York, 1976, pp. 396–403, for a discussion of human needs and their motivating effect on individuals.

technical knowledge to human needs. We carefully differentiate between engineering and the technical knowledge that is so important to it.

In engineering practice the term *technical knowledge* means for the most part *physical knowledge* — physical science, the understanding of physical materials, and the methods of applying them. We do not, however, limit engineering thinking to physical science alone, because then human needs would be neglected.

You can define engineering, then, as *the application of technical knowledge to better meet human needs.* This is a specific definition to which we can return when the details of practice need straightening out. Remember that intermediate products or provisions — such as steel mills, transportation systems, defense, hospitals — are not ends in themselves, but are of value only as they contribute to the satisfaction of real human needs.

Your Own Personal Profit Motive

Right here at the beginning, we should take a minute to tie together engineers' personal interests with their professional interests. Recently, an engineering senior asked a question to this effect: "The purpose of engineering is to serve the needs of others. So isn't there something wrong with my wanting to make money in engineering?" Of course not! Wanting to earn a good living for yourself and your family — meeting your own need for food, clothing, and shelter — is not only alright, it is essential. In our free-enterprise society, people receive the most for themselves when they are contributing the most to the general welfare. To some degree, but not perfectly, success in engineering can be measured by your own honest prosperity.

You may find at some point in your career that you are not making satisfactory progress financially or in professional responsibility in your work. Then it is time to look critically at your own performance and your choice of activity. Are you *really* filling human needs? Or are you just playing with technology?

A good engineer today should be financially comfortable in any working situation, whether in consulting practice, as a responsible employee, or in business as an engineer-entrepreneur. And every member of a successful project team earns the strongest kind of fulfillment.

How do you maximize your "take" from engineering practice? How are both your earnings and your emotional satisfaction to be increased? Certainly not by the specific decisions you make for a client or an employer. All ethical professionals agree that professional decisions can never be made in the light of the practitioner's own financial interests. It simply violates the definition of engineering we have developed above. The recipe for increasing one's return of all kinds is to look back at the definition of engineering. There is no shortcut.

The Five-Part Engineering Cycle

The engineering process involves several more or less sequential parts. For our purpose we might list them like this:

Conceive: Get new ideas.

Experiment: Try them out.

Design: Work out the details and record on paper.

Make: Build one or more from the design.

Test: Try out.

Recycle: Repeat and improve as needed.

The term *recycle* is added to indicate that in almost every project the previous activities on the list, or some of them, are repeated several times before a thoroughly satisfactory result is obtained. This recycling can take place anywhere, not just after testing. For instance, after the first experiments with a new idea it is usually necessary to recycle back to the conception stage for more or better ideas.

Suppose, as an example, that several engineers are associated in a company that makes washing machines. They are bringing out a new model. Start at the top of the list. Their first effort is to work out a conception of a new machine. It should offer more to their customers functionally or economically than competitors' machines. But some of their ideas may be good and others may not. They experiment — in a laboratory and on paper — to find which suggestions are feasible. Experimentation will also uncover problems and possibilities not foreseen at the start.

The results of the experiments demonstrate with some assurance the particular features that should be included in the machine. A detailed design is made, and a model is built and tested. The results of tests will usually indicate desirable design changes and may even suggest new concepts to be tried experimentally and incorporated in a new design. Thus, the whole cycle, or parts of it, is repeated until a reasonably optimum machine is produced that must *perform better* than the models it replaces. Our definition of engineering is satisfied.

It is interesting to note how the five parts of this engineering process are interwoven with the total environment of the washing machine. Conception, to be at all realistic, must consider the end use and the user as well as the technical state-of-the-art which allows for improvements. In addition to laboratory experiments and paper calculations, the engineer's experimental work may take the form of user surveys or observations of people washing clothes to determine the limitations of earlier models. The design must remain within technical limits of physical possibility, and the machine should be easy to manufacture. It must also be aimed directly at the function the user wishes to perform, personal convenience and tastes, the capabilities of the average repairman, the budget limitations of the average purchaser, the amount of space usually allotted for this function in contemporary house design, and similar factors.

Testing, far from being an isolated technical function, must be as realistic as possible, often including trials by typical users. Ultimate testing is use of the product as manufactured, its sales record, its maintenance record, and its reputation. Thus, every part of engineering activity (that is, application of technology to better fill some human need) is intimately and inextricably associated with the human environment.

Such consumer products provide obvious examples of this association. But all other engineering work will be seen, on careful examination, to have a relation to its environment that is no less close. The long-span bridge, the military aircraft, and the four-lane highway are designed to function in their complex environments. The machine tool, or industrial equipment in general, must operate effectively in no less complex environments.

Our five-part list of the engineering process has another interesting characteristic. All five parts apply to each of the steps individually. The experiment must be tentatively conceived and tried, designed, carried out, and evaluated. It will surely be recycled one or more times to bring out what is actually sought. Again, an economical and effective factory test procedure must be conceived, evaluated, designed, built, tried out, re-evaluated, and adjusted.

Engineering Is Creative

An important fact about engineering is that it looks to an improved present and future rather than a mere continuation of the past; hence, the word *better* in the definition of engineering above. One does not improve the supply of goods to meet human needs by doing things the way they have always been done.

Engineering inherently means change. It means improved methods of manufacturing housing materials, for example — a way to get more and better materials for less human effort. It could mean an improved way to transport food grains, in which movement and distribution require fewer work-hours. Engineering seeks continually to provide *more good* things with *less effort* to more people. You will translate this term *good* in different ways in engineering work — more attractive, more durable, disposable, recyclable, less expensive, more useful, simpler, easier to design, harder, softer, easier to train operators for, more flexible, less environmentally damaging, and so on, for an endless list.

Thus, a primary characteristic of engineering is that it is *creative*. If your work is not creative, is it really contributing to better meeting human needs? If it is not creative, it is not engineering. It may be good maintenance, good factory administration, or good something else. But it is not engineering.

Creativity for its own sake may be useless. The world needs purposeful, problem-solving creativity that carries all the way through to an attained goal. It may be a new method of welding ships or an improved device for receiving radio broadcasts.

Chapter 18 will show that creativity can take many different forms. *Creativity is the ability to combine usefully existing things in new ways.* For example, an engineer with imagination may find a way to put a certain type of plant in a difficult location that has never been tried before. This may be possible through a new combination of process, transportation, power supply, and operating schedule — none of which are new in themselves. Engineers make every effort to minimize the routine and emphasize the creative.

Engineering Rests on an Extensive Technological Base

The sum total of technical information that exists at the time any particular engineering project is carried out can be considered a platform upon which the new project work is erected — a technological base. This technological base is unusual in two respects: its unbelievable vastness, and the fantastic rate at which it is changing and increasing. Most of it is stored in millions of books, papers, drawings, and especially technical journals. All this material is scattered over the world. Some information, stored only in people's minds, is generally inaccessible.

Engineers are usually most interested in *current* and *new* technical ideas because they expect to improve and create. This new technology is found most often in journals. There are now about 100,000 different technical journals published in dozens of languages. Between one and two million technical articles are published every year. For a practicing engineer to make any significant use of this material — even limiting it to what applies to a particular field — is a staggering problem, and one which will be considered further in later chapters.

The relation of science to engineering is discussed in some detail in chapter 2. We can observe at this point that science (considered as a body of logically organized and interrelated knowledge) is a major part of the engineer's technical information. Many past engineering triumphs have come from new combinations of empirical information. Future engineering progress may be derived in great part from scientific advance.

It is a major thesis of this book that engineering is *not* only technical knowledge and that confusion of the two leads to endless difficulties and a deadening mediocrity in practice. But we must thoroughly appreciate the importance of technology to engineers. Our definition of engineering included the phrase, "application of technical knowledge" as an essential element.

Engineering Work Is Project Work

Many types of human activity seem to consist of a series of regular and predictable tasks and decisions performed on a day-to-day basis. These activities add up to useful and profitable results with little further integration or planning. Examples could be supervising or working in a routine accounting operation, operating a machine tool, or doing housework. These come to mind at once as kinds of work in which the day-to-day duties offer little variation or adventure.

But some people would disagree heartily with these conclusions and should be applauded in their dissent. We must all learn to handle our routine imaginatively, efficiently, patiently, and alertly. But it is certainly a fact that those engaged in routine occupations often find their fulfillment outside the working situation. The accountant is a service club leader; the machine operator is number one in his bowling league; and the housewife runs the garden club. To be sure, the accountant must cope with annual closing and quarterly statements, the tool operator is switched to a new machine periodically, and the housewife will prepare for school opening or closing or will move with her family to a summer cottage. But on the whole the tenor of activities is quite predictably continuous. Any occurrence at all unusual is highlighted by the very uniformity that ordinarily prevails.

Engineering is almost completely divorced from this concept of the routine and continuous. Engineering work is project work. To fulfill the defined purpose of better meeting a human need, the engineer must think in terms of a specific project, that is, a group of interrelated activities that can add up to providing for some known want. The project may be a bridge to be built, a new machine to be designed, or a new computer programming technique to be worked out and perfected. The engineer's work is a series of projects, sometimes taken one at a time and sometimes handled simultaneously. They may vary in duration from a few days to several years. Their costs can run from a few hundred dollars to millions of dollars. They may occupy only one person or may require thousands of contributors.

It is another characteristic of engineering to guide and measure project effort, and the tasks that constitute this work, in terms of three quantities

time   money   specifications

Time is money in a very real sense in this kind of work. If a project takes twice as long to complete as planned, it will, in general, cost twice as much money. Time is also important in that with the state-of-the-art in technology changing rapidly, a slow project may well be passed by new developments before it is complete. We will see in chapters 3 and 4 that coordination between different individual or group contributors to a project requires careful scheduling.

Similarly, money and money schedules (budgets) are indispensable for planning, guiding, and evaluating projects and their parts. Engineering economy textbooks often make the point that any project must justify itself by having an output that exceeds input. That is to say, if a certain economic good is to result — say ten million dollars worth of an improved washing machine — the project to design and produce it must cost less than the resulting good. In like manner, the cost of each task of a project, in terms of effort and material expended, must be less than its value to the overall project.

Specifications are a complete and precise written statement of what is to be accomplished on an engineering project or task. We will discuss specifications later in connection with projects and project control in chapters 3 and 4. Let us observe here that the engineer, like anyone else, must have a clear goal in mind in order to accomplish it effectively. But engineering tasks are complicated, so there is a need for deliberate and expert attention to specifications.

The engineer who neglects time, money, and specifications is like the banker who lends other people's money without regard to the time of repayment, the sum involved, or the interest rate and security received. The engineer is dealing with a client's or an employer's resources, plus his or her own time. In a much larger and more important sense, the engineer is handling a country's and the world's resources.

Don't Be a "Responsive" Engineer

The bold picture we have roughly sketched of engineering and the engineer may well raise some serious questions. Isn't the function of engineering more properly one of supplying technical tools that others will employ to fill human needs? Isn't it the businesspeople's function to decide what to do with a steel mill? Isn't it the farmer's function to decide whether a new tool is needed and if so, to ask for it? Isn't the engineer really an employee of the large corporation or of government? What input can the engineer possibly have to these major concerns other than the general political process?

H. G. Thuesen* develops an interesting concept of *responsive* and *creative* engineering:

"The engineer who acts in a responsive manner acts on the initiative of others. The end product of his work has been envisioned by another. Although this position leaves him relatively free from criticism, he gains this freedom at the expense of professional recognition and prestige. In many ways, he is more of a technician than a professional man. Responsive engineering is, therefore, a direct hinderance to the development of the engineering profession.

"The creative engineer, on the other hand, not only seeks to overcome physical limitations, but also initiates, proposes, and accepts responsibility for the success of projects involving human and economic factors. The general acceptance by engineers of the responsibility for seeing that engineering proposals are both technically and economically sound, and for interpreting proposals in terms of worth and cost, may be expected to promote confidence in engineering as a profession."

To be well done, professional work must be looked at as a whole. For most of us it is much easier and more comfortable to narrow our outlook to one or two parts. The national result of such a view is seen in situations such as misapplication of ideas and materials, pollution and waste, misunderstandings and mistrust between technical and nontechnical people, high costs, and aborted projects. Further, a narrow outlook limits engineering careers and personal development.

To be sure, specialization and expertness in each part of the engineering process are needed. More than most fields, engineering involves vast accumulations of physical information. There can be no such thing as a universal engineer. Nevertheless, every person must bring to bear on every small project a general and intelligent comprehension of its purpose and overall contribution. "Responsive engineer" comes very close to being a contradiction of terms.

On your first job, you may not find your associates doing this kind of

---

* H. G. Thuesen, W. J. Fabrycky, and G. J. Thuesen, *Engineering Economy,* 5th ed., copyright 1977, pp. 15, 16. Reprinted by permission of Prentice-Hall, Inc., Englewood Cliffs, New Jersey.

thinking. Few institutions on this planet have reached perfection. And the daily press of schedules and deadlines must have priority. Engineering managers will usually have enough to do in developing their beginning subordinates technically. But the wise beginner will make a systematic effort to see how his or her part of the work fits into the total project. You can observe your manager's work in fitting the project into the total company environment. By combining these practical observations and experience with wide and more general reading and study, the engineer can develop rapidly.

Operations Auxiliary to Engineering

Engineering is sometimes considered from a strict standpoint to be limited to the areas directly and immediately associated with the five-part engineering cycle. Broadly, then, people doing "engineering" are occupied with research and development, design, manufacturing, testing, and technical marketing work. The management of all these functions is, of course, a vital part of engineering work. Nothing is more unthinking than the sometimes heard remark, "He left engineering to go into management."

There are many other more or less peripheral operations which, although beyond this strict interpretation, are important and essential to modern engineering. They include, for example,

> teaching
> recruiting
> training
> industrial relations
> purchasing
> standards work
> technical advertising
> technical writing
> patent law
> engineering administration
> contract administration

Many people with engineering training and experience are occupied in these areas, particularly in their supervision. Whether or not these operations should be called *engineering* seems a matter of small importance. Their functions are necessary to the profession. A person engaged in this work can perform it with the same excellence and professional results as one doing design. Excellence in these endeavors involves an understanding of the whole engineering process and the same view of the overall goal.

The following chapters, which show how to relate each step in practice to the purpose of engineering, should be as valuable to these engineers as to others.

Suggested Readings

Readers will want to delve further into areas touched on in this book, particularly those in which their immediate interests, assignments, or difficulties lie. An almost unlimited amount of material is available. This limited list, arranged by chapter, is a good start.

It is important to keep in mind that reading and study alone cannot solve problems. Rather, it is the combination of thoughtful study with remedial *action* — trying out the new ideas in daily work — that brings results.

Chapters 1 and 2

John D. Kemper, *The Engineer and his Profession*, 2nd ed., Holt, Rinehart & Winston, New York, 1975.

Melvin Kranzberg and Carroll W. Pursell, Jr., *Technology in Western Civilization*, 2 vols., Oxford University Press, London, 1967. A good general history of engineering.

Chapters 3 and 4

Joseph J. Moder and Cecil R. Phillips, *Project Management with CPM and PERT*, Van Nostrand, New York, 1970.

Chapter 5

John B. Bennett, *Editing for Engineers*, Wiley, New York, 1970.

George C. Harwell, *Technical Communications*, Macmillan, New York, 1970.

Nell Ann Pickett and Ann A. Laster, *Writing and Reading in Technical English*, Canfield Press, San Francisco, 1970. Detailed analysis with many examples.

Chapter 6

James L. Adams, *Conceptual Blockbusting*, W. H. Freeman, San Francisco, 1974.

D. K. Carver, *Introduction to Data Processing*, Wiley, New York, 1974.

Chapter 7

Gordon L. Clegg, *The Science of Design*, Cambridge University Press, 1973.

Chapter 8

H. C. Fuchs and R. F. Steidel, Jr., *10 Cases in Engineering Design*, Longmans, London, 1973.

Peter Gasson, *Theory of Design,* Harper & Row, New York, 1973. A stimulating, thoughtful book.

Chapter 9

Elwood S. Buffa, *Basic Production Management,* 2nd ed., Wiley, New York, 1975.

Chapter 12

Philip E. Hicks, *Introduction to Industrial Engineering and Management Services,* McGraw-Hill, New York, 1977. From chapter 7 on, this book is particularly illuminating for systems.

Donald R. Plane and Gary A. Kochenberger, *Operations Research for Managerial Decision,* Irwin, Homewood, Ill., 1972.

Chapter 15

Herbert G. Hicks and C. Ray Gullett, *The Management of Organizations,* 3rd ed., McGraw-Hill, New York, 1976. The most practical extensions of this chapter will be found in the human motivation chapters of basic management books. This book is particularly clear and interesting.

Chapter 17

C. Dean Newnan, *Engineer-In-Training License Review,* 8th ed., Engineering Consultants (P.O. Box 6701, San Jose, Ca. 95150), 1976.

William E. Wickendon (and subsequent ECPD Editors), *A Professional Guide for Young Engineers,* The Engineering Council for Professional Development, New York, 1967.

Robert S. Rosefsky, *Getting Free,* The New York Times Book Company, New York, 1977. Some interesting thoughts on starting your own business.

Chapter 18

Edward deBono, *Lateral Thinking: Creativity Step by Step,* Harper & Row, New York, 1973.

George M. Price, *The Practice of Creativity,* Harper & Row, New York, 1970.

Summary

1. Engineering is the application of technical knowledge to better meet human needs. It is not technical knowledge. It is a doing. The engineer's technical knowledge is primarily an understanding of physical science, materials, and devices, but it is not closely limited to this area.
2. Conceive, experiment, design, make, test. All engineering projects go through this five-step process, and almost all repeat or recycle it (or parts of it) several times.
3. Other professions and activities share with engineering its goal of sup-

plying human needs. There is much common ground and overlap among them. Engineering is differentiated from the others primarily in that it accomplishes its purpose by the application of technical knowledge, mainly in the area of physical science.

4. Engineering is characterized particularly by the facts that it must be creative, it rests on a base of technical knowledge, and it is conducted as a series of projects.

5. Time, money, and specifications are the terms in which the engineering project is conceived, controlled, and evaluated.

6. Specialization is needed because of the vastness and complexity of today's technology, but engineering cannot be fragmented in its total process of better satisfying human needs. Each engineer must consider the project as a whole if his or her own part of the work is to be effective and creative.

Chapter Two

The Engineer

Are Engineers People?

When you think about rising in your profession, about getting ahead—that is, when you think about doing better engineering work, designing more fluently, selling more effectively, managing more productively—where do you start?

Although there is only one correct answer that you can give to this question, let's look at a few other possibilities first. I think that the majority of young engineers would say something like this: "Well, I'd see if I couldn't get some more training in machine design. That's what I'm doing now." Nine out of ten electrical engineers would say, "When I was in school we had some transistor courses, but we didn't go far enough. You know that field is changing fast. What I need is a course on the fundamentals of solid state. I'm going back to school for some more physics." That's the kind of answer I gave too, shortly after getting out of engineering school. Others will surely say, "Everything is going onto computers these days. You can't get ahead unless you understand more about them." Also, it's becoming popular now to give this kind of answer: "Technical training isn't much good unless you can apply it in business situations. I'm going to work nights on a master's in business administration."

What's wrong with these answers? Nothing, really, as far as they go. In fact, they show real alertness to the problem of our rapidly changing and expanding technological base. Most engineers that I know today are keenly aware of the need for continually updating themselves technically. There is no question but that, from a *short-range* viewpoint, what you and I most need, coming out of school, is a stronger grasp on technology.

To me, the real fault in these answers is that they fall just short of the main point: a point that engineers find hard to think about. Now I can hear you saying, "Out with it. What do you think the answer should be, Mr. Author?"

14

When you begin to be interested in getting ahead, where do you start? *With yourself.*

And you protest, "But that's what those four answers *said* in so many words!"

Well, not really. Suppose that you are on a project team working to improve the performance of internal combustion engines. You have an engine on the stand that has fallen down in torque output and is running roughly. Where do you start? Someone says, "Adjust the mixture." Another suggests, "Use a finer fuel filter," and so on. But you, as an experienced engineer, say, "Let's find out exactly how this thing works. Then we can easily make some measurements and see what it needs."

That's the point about improving your own performance. You have to start with yourself. Find out how the thing works first.

You object again, "The 'thing' you're talking about is people. Nobody knows how they work."

Well, maybe you're right. Nobody does know—at least, nobody knows the whole story or anything near it. But since you have to deal continually with people, *and especially yourself,* you may as well start to find out all you can.

There is a funny thing about engineers—when they choose this profession it is because of a tremendous interest in technology, and properly so. In their early schooling they display an aptitude for mathematics and science. They like to work with *things*. Their world revolves around devices, structures, ideas, how-to-do-it. And this is good. Heaven help the profession if people were coming into it who weren't deeply interested in technical things.

But the funny thing that we mentioned before is this—because the embryo engineer grows up with this kind of thinking, he tends to show relatively little interest in people. Frequently, he doesn't find out too much about them. Eventually, though, he has to wake up to the fact that people are his customers. It is their needs he will serve (and serve better) if he is really to engineer at all.

Now the payoff of this more or less "peopleless" attitude that all we engineers have—it's exactly the way I felt, too, as a beginner—is that we are almost incapable of seeing ourselves as persons. We take ourselves so much for granted that in effect we have drawn a little circle around us. *All* of our thinking is devoted to things outside the circle, and mostly to physical things at that. When the average engineer wants to improve his performance, it never occurs to him to find out how "the thing" (himself) works first. Look again at those typical answers at the beginning of the chapter.

Engineering colleges don't help much here either. Collegiate engineering curricula build the technological background that every engineer needs. The part of the curriculum not pre-empted by physical technology and mathematics is used for essential nonengineering subjects like English and economics. A very few culturally broadening courses in literature, history, or languages may be included. Even here, a man's main interest tends to minimize what might be gained from these other fields.*

In many engineering colleges little or nothing is taught on engineering practice as such. Design courses which were popular twenty-five or thirty-five years ago have largely vanished, casualties of a fantastic increase in the amount of technology. Because more technical information is available, we assume that it should be passed on to the engineering student. However, some hope that the way in which courses are taught will instill useful approaches and attitudes. (A few schools have developed excellent engineering analysis courses in which whole problems or well-simulated whole problems are considered.†)

Thus the mental picture you can draw of an engineer "finishing" his education depicts a man who is well trained in the basic technical knowledge of his field, especially if he has stayed for a fifth year and an M.S. degree. He has not been specifically trained in engineering practice. In fact, because he gives so much interest, effort, and time to mastering basic technology, his broader human education is rather limited.

Now it is this *human education* which will sustain and strengthen you in your work with others. It even teaches you to understand yourself to some extent. Therefore you will have to amplify it on your own, using the same drive and intelligence which made you successful in technical education. (There will be more on this topic in Chapter 17, after we have looked at some of the details of practice.)

Is Engineering a Trade or a Profession?

There have been engineers for a long time, many thousands of years. As long as man has been consciously trying to change and improve things or to build extensive works, he has been engineering. Ancient aqueducts, irrigation systems, large buildings, land survey, fortifications, roads, bridges—the work to produce all these fits well into our definition

---

* At a party recently a young lady who was studying to be a teacher—of English, I believe—complained to me about her dates with engineering students. "They are so dull," she said. "They seem to have nothing to talk about."
† See Chapter 6.

of engineering: intelligently applying technical knowledge to better meet human need.* In the past, technical knowledge was almost entirely empirical. Much of it is still empirical today, although contributions of a mushrooming science underlie the almost unbelievable engineering achievements of the last fifty years. (Confusing science and engineering leads to trouble, as we will see shortly.)

Historically engineers have been trained by apprenticeship. Until about one hundred years ago there were no engineering schools. In those days the beginner associated with practicing engineers, assisting them with detail work and minor tasks. Slowly but practically he learned his trade or profession. Without the advantage of training in a formal school, the apprentice mastered fundamentals as well as practice on the job. Of course, there was much less technology to be learned in those days.

In the mid-eighteenth century the Duke of Bridgewater wished to build a canal from his coal mines to Manchester to reduce the high cost of freight. "He called upon a local mechanic virtually without education. . . . All that James Brindley knew about canals when he started was what he had heard from the Duke; before he finished, Brindley was a celebrated engineer."†

We engineers today are still trained mostly by apprenticeship *on the job*. After a thorough four or five years of basic instruction in technical fundamentals at school, the novice associates himself with experienced engineers. By actual work on the job he begins to learn how to practice engineering.

Progress in learning is an individual matter and varies from one person to another. Some seem to learn as much and to progress as far in the first year or two as others do in the first ten. Although there are exceptions, for many engineers academic standing in engineering school has little relation to rate of development on the job. When you go out into practice you are literally starting all over again.

Does today's engineering graduate enter a trade or a profession? The answer would seem to be largely up to the individual. Historically the man who worked into engineering as described above was *first* a tradesman in some field of skilled labor. Excellence and drive singled him out for more important tasks. Eventually he had enough successful experience behind him and a good enough perception of needs to work

---

* There are many interesting books recording past triumphs of engineering, for example, *Engineering in History*, by Richard S. Kirby, Sidney Withington, Arthur B. Darling, and Frederick G. Kilgour, McGraw-Hill, New York, 1956.
† *Ibid.*, p. 208.

on his own. With sound ideas to meet these needs, he could sell the idea that he himself should be given a commission to build a new canal, to construct necessary buildings, to drain some swamp, or to provide a cross-country signalling system.

From this historical development it is easy to see that there is no precise line between the skilled trades (technician's work) and engineering. You can't tell where one starts and the other leaves off. Their centers of gravity, however, are clearly distinct, that is, there is no question of the difference between technician's work and professional engineering work as a whole.

The engineering technician today, like his predecessors of yesterday. is a skilled tradesman. He does an important job in applying technology (usually from one somewhat *limited* field) to *limited* aspects of the engineering problem. For example, he may be an electronics laboratory expert who builds and tests devices conceived mostly by an engineer. He may be a programming expert, translating problems into computer language and running them for solutions, or he may be trained and experienced in any one of dozens of other fields. He needs both specialized training and practical experience to do his job well. This training frequently requires two or more years of college work.

Today's engineer is a professional with *wide* fundamental technical training, practical experience, and the inclination and ability to *broadly* consider problems involving the *application* of technology to human need. He and his associates solve them completely through to an ultimate conclusion. Though he will probably have specialized in one or more areas of competence, his breadth of technological training and outlook fit him to take the responsibility for contributing extensively to a complete solution of the application problem.

A famous study of the American Society for Engineering Education came to this conclusion: "The engineer of the future will be expected to meet even greater challenges." He will have to be "prepared to exert positive leadership in accepting larger social responsibilities."*

Technology has grown to a scope and depth which require extensive specialized college training. We no longer find it practical and efficient to begin engineer selection from outstanding performance in a trade. Today the potential engineer commits himself to at least four years of intense study of technology before he can try his hand at any practical part of engineering. He must show considerable aptitude for abstract thought before he can even begin his years of formal study.

* American Society for Engineering Education, *Goals of Engineering Education, Preliminary Report*, October 1965, p. 21.

Scientists and Engineers Are Different

The American public seems hopelessly confused about science and engineering. Some engineers and scientists share this confusion with their ultimate customers. Since the difference between these two important activities is fundamental to our work as engineers, let's look carefully at both.

We have seen in Chapter 1 that *engineering* as an activity must be distinguished from "engineering" as a body of information and techniques. We refer more properly to this latter concept as "technology" rather than as engineering. Thus the practice of engineering rests firmly on a large and diverse technology.

Similarly *science* can be thought of as an activity or profession to be practiced and as resting on "science" as a body of knowledge. This latter use of the word is quite common. For the purposes of this chapter, however, let us treat science as an activity which also rests on technology. *

We have seen that the purpose of the engineer is to intelligently and creatively apply technology to *meet human need*. This is not the purpose of the scientist!

The scientist works to clarify and extend human knowledge, that is, to add to and clarify technology. He works for the purpose of obtaining a better understanding of the nature or environment under investigation. Ideally he would put all of the empiricism of 5000 years of engineering practice into a logical, coherent framework of scientific understanding. He works to know why and how things happen. His purpose is knowledge *for its own sake*. This is not the purpose of the engineer.

Since engineers and scientists share a common technology and work with and on the same physical environment, their activities touch and overlap at many points. Nevertheless the work of scientists and engineers is clearly differentiated in terms of purpose and goals. It is difficult to see how either can do good work in his field while confusing his objective with that of the other.

No doubt most scientists hope and expect that their work will be of ultimate value to humanity. But it is not performed *directly* with that as a goal.† Most engineers will at one time or another be adding

---

* Some scientists will object to including scientific knowledge with technology on the ground that technology includes a great amount of relatively minor detail and empirical art in addition to "true" scientific knowledge. It will be seen that this point is not important in the present discussion.

† An interesting paradox about scientists says that they will produce more useful ideas if you just ask them for ideas rather than ask them for useful ideas.

to the store of technological information with which they work—but for the specific purpose of meeting a human need. In a sense this is a by-product, although an important one. It is not technology for technology's sake.

There is no reason why a man trained as a scientist cannot work as an engineer, or vice versa, though it would seem somewhat unusual for an engineer to set aside his practical problem-solving urge for long in order to pursue a dedicated search for knowledge itself. In fact, some men appear to work successfully as both engineers and scientists. This can be done with real success only by carefully differentiating between the two purposes, by wearing one hat at a time.

A good example of this boundary crossing occurred in the development of radar during World War II. Not enough engineers were familiar with microwave techniques and their physical principles to carry on the work. Instead it was successfully guided (and to a large extent performed) by physicists and others trained in the abstract sciences. These men were not acting as scientists in this development but mostly as very practical, problem-solving engineers and engineering managers.

The term "engineering science" is heard more and more of late. It refers to neither engineering nor science, as we have used the terms here, but to the systematic part of the technology on which engineering is based.

We engineers owe a great deal to the scientists and always will. Of course, this help works in both directions. Modern science relies very heavily on engineering works and devices for its continued extension. Many scientific investigations depend on the excellent engineering in modern oscilloscopes. Space programs, which are permitting our scientists a close look at other planets, could not advance without the phenomenal engineering achievements that make space travel and space probes possible.*

Over the past few centuries (particularly since about 1850) technology has been enriched and strengthened by modern science. Much empirical knowledge has been clarified and put on a solid basis of understanding. New physical knowledge has added to what already existed. These effects of science result in improvements to old things. For example, bridges can be built with less material because material stresses in design are now quite well understood. Even more important, however, is the fact that scientific advances in technology make possible entirely new

---

* "In fact, engineering contributed far more to science than science did to engineering until the latter half of the 19th century. . . . This is not surprising when one remembers that the empirical knowledge of engineering had been accumulating for at least 25 centuries when the Greeks began their scientific investigation of nature in the 6th century B.C." (Kirby *et al.*, *op. cit.*, p. 43).

devices and structures. A brand new electrical communication is in use because electricity and magnetism have become a "science."

There is another striking but not untypical example of *both* improved devices and new ones from a single major scientific advance. Although dry-type rectifiers had been used for decades in power and instrumentation applications, nobody really understood how they worked. Design was empirical. Research efforts to understand what was taking place many years later led to a satisfactory theory of rectification through semiconduction. With this understanding improved rectifiers could be designed, but results of this successful research did not stop there. New understanding led almost directly to transistors. Subsequent research and development over a period of ten years launched an entirely new industry.

Engineering Specialties

Every engineer is both a specialist and a generalist. Some specialize to a greater degree than others. But whatever your specialty is, you must always think of your daily work in terms of ultimate application and benefit. To do less is to throw away the guidance you need to effectively and creatively apply yourself.

There are two kinds of specialization. The first comes about because technology is too vast for anyone to be expert in even one broad area of it. Besides being a mechanical, or electrical, or some other kind of engineer, each man comes by practice and study to be especially competent in a specialty like high-frequency circuits or bearing design or the application of reactor control. Note carefully that this kind of specialization applies to *technology* rather than to engineering practice. The specialist in this sense is a complete practitioner of engineering, but he limits the *technical area* of his practice to certain specialties. It would probably be more accurate to say that in his practice he emphasizes certain specialties.

A second kind of specialist emphasizes one or two of the five engineering functions described in Chapter 1. This specialization is especially common in larger companies. Some engineers will be concerned with advanced development work, for example, whereas others may be responsible for specific parts of the production or building effort.

In this second type of specialization it is essential not to lose the big picture of engineering's ultimate purpose. For example, the advanced development man thinks to some extent about the design and production problems of what he is working on. He looks at his product frequently from the user's point of view. Similarly designers cannot afford to forget

᷉r production and use or conception and experimenting. They look to the conceivers and experimenters for their next new ideas.

A man not specialized in any way usually has little to offer the world. Conversely, the man whose specialty blocks out most general considerations in engineering practice will find his usefulness limited. Only a very few *narrow* specialists can be used; most of these are employed in large companies.

"How much and in what area shall I specialize?" is thus an important question for the practicing engineer. A good rule is to make two efforts in personal development—one aimed at developing, maintaining, and expanding a special competence in one or possibly two areas, the other at keeping thoroughly abreast of the profession as a whole and its major developments.

Although I have encountered a few engineers whose usefulness was somewhat limited in that they had almost no special competence, over-specialized individuals are much more common. They are overspecialized, not in the sense that they know too much about their specialty, but in the sense that they have neglected their general background and professional outlook. They cannot effectively apply their specialty to everyday problems.

"All right," you say, "so I'm going to specialize and still be careful not to drop the broad outlook. Now how do I choose a specialty?"

For most of us this is a mysterious process that works out unusually well. Older engineers in looking back at their early careers will say, "I guess I was just lucky. The department needed someone to go strong in foundations, and I'd happened to finish that part of the graduate work. Somehow it just worked out."

Actually specialties work out for most people as combinations of strong interests and organization needs. When you find something unusually interesting in your work, take odd moments to dig in a little. You may be developing a useful specialty.

One of the most successful specialists I know is an engineer who devotes perhaps 95% of his time to development work on special switching circuits. These combine solid state and saturable core devices. Yet this engineer thinks habitually in terms of industry needs and applications and zealously guards his company's patent position. He takes the lead in planning and conducting educational sessions for designers who will use his ideas; he works with very practical attention to production costs. And he makes it his business to prod his managers if they are neglecting the broad implications of his specialty to the company.

This man worked out his present specialty from previous specialization in magnetic amplifiers. All specialties are transient things at best; the more narrow they are, the shorter duration they will probably have.

This is another serious reason for not neglecting the broad aspects of general practice. You will probably have a number of specialties in succession during your career. There is nothing sadder than a middle-aged engineer who has made genuine contributions in an earlier specialty but now finds himself unable to change with the times.

Computers versus Engineers

We read in the papers that computers are taking over more and more design work from engineers. That sentence should probably be rewritten: "computers are taking over more and more so-called engineering and design work from men who have graduated from engineering school." From the definition of engineering developed in Chapter 1 you must agree that a computer cannot practice engineering. By the same token a man who is devoting any substantial portion of his time to the repetitive and programmable tasks which can be relegated to a computer is not practicing engineering either.

As an interesting example consider a project of one large company to automate the design of a series of instruments. In a special order department they manufactured thousands of different meters by combining a few hundred variable characteristics in various ways. For example, they had round cases and square cases, perhaps three or four standard sizes of each. They had d-c instruments and rectifier instruments. They had perhaps ten different voltage ranges of coils. There were left-hand scales, right-hand scales, zero-center scales. There were different colors and finishes. So their special line consisted of an almost endless number of combinations of a few simple variables.

When an order came in, someone had to sit down and interpret the customer's requirements, translate them into appropriate range specifications in the line, consider whether all the factors involved were compatible, draw out the design, place material orders, schedule the work in the factory, give the customer a promise date, work out cost and pricing, and then follow the order through to an ultimate shipping event.

Not surprisingly, the company discovered that a computer could do all this work for 95% of the orders; and it could do it better, quicker, and cheaper. The customer order was transcribed onto a punched card and fed into the computer. From then on the computer took over, "designed" the instrument, checked the compatibility of requirements, placed an order on the production group, scheduled production, ordered the material, wrote the customer a promise date, and priced the instrument. In fact, on the side the computer could maintain inventory records and statistical sales data and could forecast future business. For the exceptional case when a set of customer requirements did not fall within the competence of a computer program, the computer itself could make this decision, inform the operator, and tell him why.

Now this beautiful and useful result did not take any engineering work away from an engineer, for we have seen that it is the engineer's business to find better ways to meet human needs. Repetitive activities do not do this. Work that can be sufficiently foreseen to be programmed for a computer is not innovative or creative. The engineer continually seeks new solutions, different ways, improved results, better combinations. The project work of devising the computer solution was in itself a wonderful example of engineering. How could that type of activity be compared to the routine work it displaced?

Thus, in your engineering practice, you will continually seek new ways to solve problems. You will make deliberate efforts to be creative. You will frequently examine your work and working habits to minimize routine, to increase professional content.

We will examine in Chapter 18 some detailed aspects of this search for personal creativity in engineering work.

Summary

1. New engineers are strong in technological knowledge but must learn more about people (and themselves) in order to use what they know.
2. The work of engineers and technicians differs in (a) the breadth of the engineer's interest as opposed to the more limited nature of the technician's, and (b) the engineer's activity with people as customers whose needs his work is filling.
3. Engineering work and scientific work are easily distinguished in terms of their goals—the one to solve problems of human need, the other to extend knowledge. Although both professions are based on technology, to confuse them leads to poor performance in either.
4. Specialization is necessary but cannot be allowed to obscure the engineer's broad view of his projects. The wise engineer makes two parallel self-development efforts—one to extend and deepen his specialty, the other to keep abreast of his profession. An engineer must plan to change to new areas of specialization as technological times change.
5. The engineer is engaged in finding new ways, improved methods, to accomplish his goals. He will make deliberate efforts to be creative He minimizes the routine content of his work as far as possible.
6. An engineering career includes a life-long dedication of oneself to continued learning and self-development.

Chapter Three

The Project
and the Project Team

This chapter looks at project teams and their performance in some detail, highlights factors which make for excellence in the individual performance of team members, and points out pitfalls which may trap the unwary.

It is a truism that today's engineering is done mostly by groups of engineers working closely together instead of by individuals. Many of today's projects are large and complex—space projects, for example. Most require the contributions of different kinds of professionally trained people. Also, today there are more engineers to be organized and managed. Therefore, extensive team effort will be with the profession for a long time to come.

The Major Factor In Project Success

There is an almost unbelievable difference in output from one project team to another. A group on one project shines as an example of success and professional progress. An almost identical group on another job is the epitome of everything mediocre and humdrum with respect to both work performance and personal development. A third group, having spent considerable money and time with substantially invisible results, and often wrangling among themselves, will find itself dissolved or completely reorganized.

We would expect some variation between groups of engineers, as we would between individuals, but not as much as actually occurs. People on two projects are trained in much the same way technically. Various types and degrees of talent are distributed in each group. Each manager organizes and supports his group with a sincere intention of getting the job done. Certainly one group seems to work as hard as the other if not as effectively.

Why then is there such wide variation in group output?* The usual explanation is that one group has a project it can solve. The other has been given a project which is too difficult, is beyond the state of the art, or is otherwise unreasonable. But this is an unsound diagnosis. Difficult problems are the backbone of engineering work.

The excellent group, faced with what appears to be an impossible assignment, finds this out quickly. It reorients itself to whatever opportunity exists, does what it can with this altered view, and goes on to the next problem. There must then be some explanation more fundamental than problem difficulty.

The major difference between excellent and mediocre groups lies in the answer to this question: *how effectively are the group members able to work together?* Other things being reasonably equal, as they usually are, the ability of group members to operate as a team, to work effectively toward a shared goal, is the real basis of success in today's engineering effort.

A close look at these two groups shows contrasting symptoms long before the end of the project:

EXCELLENT GROUP	POORER GROUP
1. Technical work generally on time and within budget.	1. Seem habitually behind on assigned task and money.
2. Engineers busy but give impression of being on top.	2. Always rushed to meet deadlines.
3. Necessary changes and redirections are made in a timely way and taken in stride.	3. Radical changes are made too late, often at the last minute, with traumatic consequences.
4. Each group member is obviously growing fast in experience, and readiness for a bigger assignment.	4. Group members feel frustrated and stagnant and complain that they are learning little.

You may say that one group has a good boss or leader and the other hasn't. This is probably true. However, it in no way invalidates our conclusion about team work, for the most important thing a leader has to do is make his team as a whole effective.

Group members themselves must know or be taught how to participate in group action. A leader trains and develops them along this line as his project progresses. Incidentally, the ability of a group member to

---

* This phenomenon has been noted in other fields besides engineering. For example, in his graduating address to the West Point class of 1869, General W. T. Sherman (the victor at Atlanta, who said, "War is hell!") observed, "The soldier in the ranks is not a block of wood or a mere unit. . . . As one man varies from another, so bodies of men vary still more. . . ."

later assume the role of project leader on a new job is derived from this conscious group experience.

In smaller companies or smaller engineering departments there is still a very significant amount of "one-man" project engineering. Some of the best engineering is done in this way. These engineers have some real advantages over those (most of us) who will work in groups, but there are disadvantages, too. Even in a simple one-man situation, however, the engineer works closely with others—with sales people and manufacturing people, for example—in providing for human needs. And he will "manage his own work." Thus the philosophy and some of the detailed techniques of this chapter are still important to him.

Why Projects?

You saw the importance of the project concept itself at the end of Chapter 1. The engineer conducts his daily work on a project basis because he is making specific betterments. True, he may anticipate forty years of similar work, but it is difficult to avoid the routine and mediocre if work is thought of as a more or less uniform, continuing process. Instead the engineer articulates his effort into a series of projects *with specific goals and time and money bounds*. In addition to focusing his attention and effort, the project concept gives advantages in ease of schedule and financial control. Details of this control will be left for the next chapter.

There are engineering positions which seem to cry aloud for the incumbents to consider them as routine, day-by-day, continuing work. A man may be responsible for all the testing in a certain company. Another may be responsible for the specification and source development of components for their products. A third may be placed in charge of maintenance for several states. A fourth may be assigned full time as liaison engineer between a production group at one location and a central engineering organization at another. Can these kinds of responsibilities be viewed in any other than a routine way? They must! Each engineer must separate out the routine and provide for it to be taken care of quickly, efficiently, and accurately. Then he will be free to get on with his really professional engineering work—deliberately and specifically working to better accomplish his responsibilities, to develop new and improved ways of testing, to effectively integrate his efforts into the projects he is representing, to creatively overcome the technical limitations that he finds on every side of him.

There is, of course, an overriding requirement in all cases to *get the job done*. Hence routine details cannot ever be neglected. But in providing for them the wise engineer will make maximum use of supporting personnel and services—technicians, clerical people, computers. (If he has no such support, of course, he will have to take care of them himself.) By utilizing "economy of force" measures here, he can then devote himself to the most difficult, interesting, and important part of the job.

A Typical Small Project

It will be convenient for us to use a specific engineering situation to discuss some of the details of project engineering. We can use it later for an example of cost and schedule control. (The situation chosen here is a simplified composite of several actual projects.)

Suppose that you are associated with a company that has been building specialized tape recorders for a number of years. Your company is considering making a recorder for space vehicles. It is to store radar data taken on a fly-by mission past a planet over a 30-minute period and then to permit slow read-out of these data for transmission to earth on a narrowband channel over a 60-day period. A large aero-space company with which you have done business before has invited you to bid for a contract to develop such a recorder for its use.

It is decided to put together a small team to take a look at the problem and arrive at one of two results—either show why it would not be a good opportunity for the company, or produce a technical proposal and bid to undertake the development.

Your manager selects you as project engineer for this effort (your first opportunity to act in this capacity), presumably because you have now had several years' good experience in recorder engineering. You and he sit down and work out the following *organization* for the job:

1. Project engineer   Yourself
2. Tape-handling mechanism   Mechanical engineer
3. Packaging of electronics and   Mechanical engineer
   over-all configuration
4. Circuitry for radar data con-   Electrical engineer
   version and recording
5. Circuitry for read-out and   Electrical engineer
   synchronization with trans-
   mitter
6. Overall noise problem, error   Applied mathematician (half-time)
   rates

You decide that in addition to your project engineering work you can also handle, on so small a job, the second mechanical engineer's duties. Three other engineers and the applied mathematician are assigned to you. Work commences.

This example could have been selected from a hundred other fields. Is a new consumer product to be designed? Is there a nagging, recurring problem on the production line that needs basic solution? Any engineering problem larger than one or two men can solve will be organized similarly.

The first question is whether the company should propose a solution and bid on the job. That is, what, if anything, can it profitably (usefully and efficiently) contribute to meeting this need? When a company decides to undertake an engineering project it has already *planned* in general *what is to be done*. In the example above it was decided to examine the technical possibilities and (if they were favorable) to propose a development effort.

After this broad planning and decision the project is organized. *Organizing is deciding how the work will be broken down and who, by name, will do it.*

This basic concept of project organization is simple. A leader, the project engineer, is chosen and is made responsible for the entire project. He and his manager divide the work into task areas which can be handled by individuals, who are then selected and assigned. Usually these tasks will call for several different types of training—perhaps a circuit designer for one, a mechanical engineer for a second, some other specialist for a third. As we will see in a moment, this task area breakdown is not meant to rigidly compartmentalize the work, but it is important in showing the kinds of people needed.

After organization comes the *doing* phase. Here the project engineer divides the daily work as the project progresses and assigns it among the individuals involved, including himself, thus completing the project as planned. We are going to see, however, that the doing phase is by no means so simple as this sounds.

Organization itself is somewhat more complicated than the above discussion indicates, since it varies with time. Some contributors are needed at the beginning of a project but not at the end. Also, some people's work will taper off. Other specialists cannot be used at all until after the project is well begun. Some contributors will work full time on the project while they are assigned to it, an arrangement that is very desirable if a man is to put his best efforts into any task. But other contributors, particularly some of the more specialized, will have to be used on a part-time basis. The need for them may develop in this way. More often, some specialists will be available for only part-time work or as consultants.

Sometimes a manager will keep the project engineer's role for himself, but this is not often successful where he has a number of projects and other responsibilities to distract him. A simple project organization makes *one man responsible* on an essentially *full-time* basis for the day-to-day operation and the total result. This is a powerful idea both to accomplish engineering work and to develop engineers.

Day-by-Day Project Operation

Now that a good project team has been organized, how is it going to operate? The daily functioning of the group separates the sheep from the goats among engineers. Here we begin to see differences between the outstandingly successful group and others.

First it will be useful to observe some complex relationships in even so simple a project as our example. There were five professional contributors. Any two of them have important parts of their work in common—areas where the detailed decisions of one will seriously affect the work of the other.

Consider, for example, the two electrical engineers. How the first contrives to encode original radar data and record it on tape will be vitally important to the second, who must devise an effective read-out system and synchronize it with a transmitter at a very low data rate. The work of both these engineers bears heavily on that of the mathematical analyst and vice versa. Each of them is obviously concerned again with the characteristics of the tape-drive mechanism and the extent to which packaging will be limited by space-vehicle environment, power supply characteristics, and so forth.

Interfaces

It will be seen that there are ten combinations of these five professionals taken two at a time. Thus there are ten technical *interfaces* between the work of these five men, most of which are of real importance and concern. The *interface* concept is a useful one. Although the word means literally a surface regarded as the common boundary between two solids or spaces, it can be applied more generally as the boundary where almost any two things come together. The two things can be ideas, for example, the interface between law and custom. The two things can be individual technical design responsibilities, as we used the term above. The two things can be the parts of an engineering system; think, for example, of the common pneumatic and mechanical connections between two cars in a railroad train, or of the mechanical and electrical elements that come together between preamplifier and receiver in a radar system.

In a one-engineer project there are still interfaces between parts of the work—interfaces which it is well to consider carefully. But it is when the job is so large that it must be divided among several contributors that the recognition of interfaces and their requirements becomes vital.

As long as one engineer has charge of all the details of a job, it will probably be reasonably well coordinated. If he designs both a bracket and a case on which it mounts, he is unlikely to mistakenly space the mounting holes on the case differently from those on the bracket. If in a radio receiver he designs both the power supply and the amplifier, he will surely specify the supply output voltage to match the requirements of the amplifier.

Coordination of Engineers

In our earlier example, if one engineer were to design both the data-recording system and the read-out system, there should be little question that they will be completely compatible. Unfortunately, for a larger project in which responsibilities are divided among a number of engineers we cannot make these assumptions about compatibility. Instinctively the man who designs the data-recording system will select a recording mode to optimize his own work. But the engineer responsible for read-out will be equally sure that the recording mode should favor his read-out problems.

Suppose that these two come from different earlier projects and are somewhat naive about group work. The recording engineer may assume that his optimized recording system will be read by a perfectly feasible means used on his last project. Similarly, the read-out engineer may assume that his optimized read-out system will be used to read a recording mode with which he is familiar from past experience. Thus the very existence of so obvious an interface as this can be missed.

We noted earlier that the project engineer might accomplish the daily project work (the doing of the project) by simply parcelling out required tasks to his workers and intelligently following them through to· completion. Although this is the basic structure of project operation, rigid and literal adherence to it is *a major source of project failure.*

We cannot expect a project engineer to resolve ten serious interface situations on his project by placing appropriate detailed specifications on each of the tasks of his five contributors. He would have to do as much work as all of the individuals in their own areas, or even more.

Are you an engineer who expects your project leader to (a) tell you exactly what to do and then (b) resolve conflicts in your work with each of the other project contributors? You will have to find a superman to work for!

Now we see what a really effective team has to do. Each member must follow at least broadly what every other member is doing. "But," you say, "do I have to keep my nose in my neighbor's business in addition to doing my own part of the job? Isn't that a waste of manpower and engineering effort?"

Yes, you must follow the whole job well enough to understand thoroughly how your part fits in. Your neighbor's business is your business. You're on the team; it's your responsibility to make it succeed. Of course the degree of effort you put into understanding someone else's contribution varies with its importance to your own work.

Where interfaces are involved, you will want to know enough to be able to work out a solution directly with the man on the other side. Then together you can take this solution to the rest of the project for the leader's blessing. Also, you will want to know enough to be sure that your neighbor is observing the interface specifications and your agreements with him. If changes are going to be necessary, the sooner you know the better.

On a poor team, members expect the project engineer to define their work specifically. They go off alone to solve problems with little reference to other contributors. After all, this was the way they learned technology in engineering school. And, frequently, on this kind of team the project engineer himself, sometimes unconsciously, feels it a point of honor to make all coordinating decisions.

On an excellent project team, by contrast, each member takes full responsibility for his own work, full responsibility that it will fit into project requirements. In practice this means especially that he participates actively in seeking out and resolving the interface difficulties. All members measure their success in terms of project success. If they accept the earlier definition of engineering from Chapter 1, what other measure of success could be used?

Project Engineer's Role

Your job as project engineer is different now, isn't it? Its mission is still the same—you take full responsibility for getting the job done on time, within the budget, and to required technical specifications.

But in the daily operation of the project you are now concerned with seeing that the interface problems are recognized and resolved by proper decisions rather than in making all these decisions yourself. You encourage free and fruitful discussion of ideas by *the whole team,* no longer limiting project management to two-man relations between yourself and each contributor. You will still have enough decisions to make, anyway.

In this new concept of successful group action each member widens his interest to the whole project, while still making his major contribution through careful and excellent work on the part assigned to him. He realizes that he can't do *his part* of the design or analysis or experimentation well unless he understands how it *fits into the whole.* He knows that when he loses sight of what the other contributors are doing (at least broadly) his own efforts can become very inefficient. A great deal of his work may have to be scrapped because it does not fit properly into the whole job.

Conditions for Effective Group Work

Let us consider an example of the effect of group members not working with each other toward one goal. In this particular project weekly meetings were being held for the ten or twelve contributors. About half of the group were from another engineering section removed by several floors from the main project. (This is often a difficult situation for a project engineer.) Each of the meetings covered well such progress as was being made—perhaps in even a little too much detail.

But three closely related areas, areas of critical importance to the project, had reported almost nothing for a number of weeks. The optical engineer couldn't proceed further until he had a mechanical layout, and he blamed the mechanical engineers for holding him up. In turn, the mechanical engineers couldn't proceed without the optical design and blamed him for delaying them. Both also partially blamed a physicist who kept insisting every week, long after the original deadline, on just a few more days' experimentation with a critical process.

Now frequent, usually regular, project meetings are one of the best techniques a project engineer has to help coordinate his people's work. From the above example, however, we can usefully draw the moral that *going through the motions* of the various coordinating techniques will not coordinate a project. It is up to all members—but especially the project engineer, who has ultimate authority and responsibility—to ensure that interface problems are in fact resolved. In the particular case considered above, the outside contributors were unreasonably expecting all resolutions to be made by the project engineer himself. He apparently thought that a simple weekly confrontation would eventually solve the difficulties, but his team was not sophisticated enough for this to happen.

In reviewing here several *good techniques* that have been successfully used in project operation we will avoid falling into the same error—thinking that, like the quadratic formula, they will automatically get results. They must be used as tools by the project engineer and his associates and used expertly. Feedback from results will tell which should be augmented and which decreased or modified, until the project is running smoothly and effectively toward its goal.

In this informal, free type of operation, it will be difficult to obtain results when project members are unable to enter wholeheartedly into the work and to identify their own success with that of the project. In an outstanding group not only the leader but also every man on the team knows what constitutes effective group action. Each man does everything he can to bring about conditions that will favor and stimulate his own work and the group work as a whole. Some of these conditions are the following:

1. Information available to any part of the group is freely shared with all.

2. In particular the current status of the project, its specifications, changes to specifications, and information about the customer or using agency are made available to all as rapidly as possible.

3. Frequent project meetings are held as needed to review (a) the project as a whole and (b) the work of individuals for progress or lack of progress. Thus changing interface problems can be recognized and resolved quickly and effectively.

4. All participants are alert to make project meetings fruitful by

   (a) digging deeply into presentations to get at the meat of situations and problems;
   (b) suppressing detailed discussion which has little bearing on the work of others;
   (c) readily investigating the possibilities of alternative solutions even if apparently unfavorable to their areas;
   (d) being ready to present at any time briefly, clearly, and impersonally the status and problems of their own work;
   (e) promptly identifying the work of other members and giving credit for it;
   (f) willingly identifying and accepting credit for their own work and contributions.

5. All members, but especially the project engineer, hedge in each area of responsibility and each interface as soon as possible with specifications (probably tentative at first). We noted earlier that working to specifications is a characteristic of engineering. The engineer is literally *unhappy* until he can get his task or project specified precisely with known techniques that he personally understands are feasible.

6. The project engineer distributes to each participant without delay memoranda of any project meeting. He includes in these memoranda brief summaries of problems discussed, major events that have occurred, and especially decisions and assignments made at the meeting. Although it is normal for the project leader to do this himself, it is a valuable experience to pass around to others at times.

7. Each member of the team watches money and time schedules like a hawk! He keeps his own time commitments zealously, but when it is going to be impossible to do so—and this will happen in engineering work, especially in research and development—he informs his project engineer and other involved colleagues as far ahead as possible.

Your Individual Problem With a Group

Although group attack on problems is standard practice in much engineering today, some real pitfalls await the engineer who works in this way. On a poorly run team one member can sometimes coast along, making a quite modest effort and contribution. A good project engineer will not put up with this for a minute. This same temptation may take a milder form: you may find yourself leaning too heavily on others. You may fall into the habit, for example, of not being really thorough in your own work, feeling that the group will have to check it out anyway. You may work passively, letting others make all the suggestions instead of practicing creativity yourself.

On occasion, a group of really first-class contributors can do an amazing job of covering for an inept project engineer. Within the project itself, however, nothing can be hidden. The poor performer or the coaster is quickly recognized by his colleagues.

Group work tends to submerge the individual. Every member must make specific efforts to minimize this disadvantage, particularly for himself. Sometimes one or two strong individuals will practically take over a project, dragging everyone else along behind them. The problem here for the manager and his project engineer is to use this energy and talent without letting it damage the project. Any serious interference with smooth and united team performance must be quickly overcome.

An engineer grows by excellently fulfilling specific, difficult responsibilities. If the project team is so poorly conducted that he has no clear responsibilities, or if he cannot identify a portion of his own contribution in the group solution, he lacks an essential job satisfaction. The engineer who cannot see some of his own professional work helping to better meet human need has not yet done engineering.

But you will find advantages in group work, too. Through team effort you can devote a larger portion of your time to a favored special area and still participate in overall project engineering (with its satisfactions and growth stimulation). Group work is natural training for potential project engineers and engineering managers. You have unparalleled opportunity to learn from more experienced colleagues through close association. Most individuals work nearer their peak capacity when they are members of a smoothly functioning team.

Thus your ability to work effectively as a team member is a major asset to your professional career. Teamwork is doubly important. First, most engineers will find themselves working as members of a group. And, second, once in a group, the engineer's *individual* morale and productivity will be heavily dependent on his standing as a team member.

An odd saying circulates about group effort, but fortunately few engineers believe it. It states that the total output of a well-run group is greater than the sum of the individual outputs of members. This is, of course, mathematical nonsense: the contributions of a project team *are* the contributions of its members. In one sense, however, the saying is true, for a well-run team stimulates its engineers. Each will produce far more in the group than he could without this mutual encouragement, practical guidance, and challenge.

Again let me warn you that the world is not perfect. Your present project group may operate with little resemblance to the hard-hitting, effective engineering team we have discussed in this chapter. Many project engineers have never seen a really effective group effort.

What can you do? Talking and complaining will not help at all. Do your part of the project as effectively as you can. Identify the interfaces in which you are involved. Sit down with the other engineer concerned, and work them out as far as you can. Then you are in a position to make practical suggestions to your leader. Nothing succeeds like success.

Good Project Team Practices

1. The project is clearly defined as a specific goal with stated resources in money, men, facilities, and time available.

2. Each member keeps informed on all major aspects of project progress and developments.
3. Each engineer takes full responsibility not only for performing his part of the work but also for seeing that it will fit into the rest of the project. He quickly identifies each interface problem between himself and a colleague and works out a solution of the problem with him.
4. The status of each element of project work is frequently reviewed by all members of the team assembled together. Interfaces are carefully analyzed and documented.
5. All team members measure their individual success in terms of project success.

Poor Project Team Practices

1. The project operates on the assumption that a clear assignment of tasks by the project engineer will adequately resolve design conflicts. Project communication is essentially limited to that between project engineer and each member.
2. Because time is always short, project meetings and the sharing of work-results between engineers are limited.
3. Project meetings degenerate into bull sessions and delve into such detail that most participants find them unproductive.
4. Some engineers on the project, fearing criticism or disliking nebulous situations, hold back their own concepts until the other work is pretty well outlined, so that their ideas are more likely to be right.

Chapter Four

Project Control

Here is a story that illustrates almost everything of importance about project control. During one of the many organizational changes that take place in any engineering establishment, I was asked to take a well-established design engineering unit into my section. These people were then doing both design and development projects. It was a pleasure to have them because a dozen of them were first class engineers in their specialties. Furthermore, the unit manager, a young engineer with outstanding experience, had a fine reputation for technical judgment.

The new arrangement was announced at the beginning of a top management review of one of this young man's development projects. My first experience with the unit occurred at this meeting.

This particular project happened to be seriously behind schedule—hence the high-level attention. The nature of the project effort had changed significantly as new technical problems unfolded with the progress of the work. Early schedules and plans became hopelessly outdated.

During the review, the unit manager, whose technical excellence was unquestioned, committed all the sins in the book. He allowed himself, his project engineer, and his project team to be put very much on the defensive in spite of their knowing more about the project and its problems than anyone else there. During the meeting important difficulties were revealed, and questions asked about their resolution. But the unit manager offered no new plans or specific remedies, dwelling rather on difficulties and imponderables. His attitude seemed to be, "Let the front office have its full emotional experience here, and then we can get back to work."

After the meeting we sat down together to get better acquainted. His first emphatic remark was, "Engineering projects simply can't be scheduled."

This experience introduces nicely several points:

1. Good technical engineering work will be obscured (and can even be completely nullified) by poor administrative control.

2. Good technical work will not by itself control a project. A project cannot control itself. It must be deliberately controlled by the project team, especially the project engineer.

3. Plan and schedule for almost every engineering project will change significantly during the work.

4. It is a surprisingly common belief among some groups of engineers that technical work cannot be effectively controlled. This attitude prevails at times even in high places. Such a conviction sometimes reflects the influence of a manager who feels the same way, or of a previous one who did.

5. The idea that projects cannot be effectively controlled is a self-fulfilling prophecy for the engineer who thinks this way.

6. From a business standpoint, company management naturally expects that technical work will be administered effectively.

What Project Control Is

Even if you have just begun your career, you should understand project control.

In Chapter 3 we considered engineering projects which required the combined efforts of a number of professionals. Most projects do. We saw that these projects, even when small, are surprisingly complex in technical relations between contributors. These relations are generally much too involved to be coordinated directly by the project engineer alone. Instead each of the contributors on an effective project team makes it his business to see that his own work is integrated into the projects as a whole (with special attention to interfaces with the work of others).

In this way, from a *technical* point of view, the project goal is accomplished in a reasonably optimum way. The team's solution does not contain unresolved conflicts between its various parts. The result will be a practical design or a useful development or a sound recommendation—whatever the object of engineering effort was.

Now it is the activity of the engineers and their use of resources which consume project funds and time, so time and money scheduling on a project cannot be separated from technical thinking and activity. It is a common mistake to try to distinguish between the two, but technical decisions are inherently time and money decisions, too.

Hence, the effort to control a project—that is, to direct it so that its technical goal will be accomplished efficiently within time and money schedules—must be shared by all members of the project team, just as technical coordination is a joint responsibility.

Control is comprised of those administrative measures taken to get a project on schedule in regard to time and money and to keep it there.

If your project is a one- or two-man job, the control effort may be considerably easier than it was for even the small project we are using as an example. But control is still needed. Although the measures used may not be so elaborate, the project still will not control itself simply through good technical work.

Again there is no question that the project engineer has full responsibility for project control, as he has for technical output. However, he needs his teammates' expert assistance in this control effort, too. Quite a bit of his own effort will be devoted to control activities. He will probably run a tighter shop with respect to schedules and budgets than with regard to technical decisions. But each of his engineers must control his own part of the work in order for the project as a whole to be controlled. Also, the engineer himself will be controlled in one way or another, so he must be familiar with control needs and procedures.

There is nothing new about all this. The ancient Egyptian engineers had projects about as big as any we undertake today—and probably a good deal longer in time span than ours. Their "inscriptions boast that some jobs were accomplished without accident or sickness and completed on time."*

The Three Parts of Control

For better understanding it is useful to divide project control into three parts:

Scheduling.
Monitoring.
Controlling.

Scheduling

Scheduling is the planning and recording part of control. Let us use the project situation described in Chapter 3 to illustrate scheduling procedure.

First the project engineer with the help of his contributors lists the tasks to be accomplished. For our project we might note the following:

1. Establish initial overall concept.
2. Determine critical factors in each area.

* Richard S. Kirby et al., *Engineering in History,* McGraw-Hill, New York, 1956, p. 22.

3. Specify all interface conditions and alternatives.
4. Establish estimates for attainable recorder characteristics.
5. Establish final configuration and make recommendations.
6. Complete proposal document and costing.

Although these six items are reasonable here as the principal project tasks, and will serve nicely to illustrate control procedures, we would need more detail in our project story to know for certain what the list should be. For an actual *development* project or *design* project, scheduling will start in a great deal more detail. This kind of exploratory proposal work is relatively simple.

How much detail should you put into your schedule? That is an important question. If too little detail is included, planning has been insufficient to foresee major problems and consequent work needed. Then the project is not really scheduled at all. Schedules provided in that case are only a delusion.

Including too much detail, on the other hand, has two drawbacks. First, it obscures the really important tasks and milestones. More important than that, too much detail tends to make solution efforts inflexible. Engineers become so absorbed in meeting schedule items that they overlook other possibilities. Also, the more detailed a schedule, the sooner it will be riven with changes.

It is always necessary that the "critical" items be considered in scheduling a project. We will think more about this term in Chapter 8. For the present take "critical" to mean those elements of the project which will have a particularly important role in determining its success or failure.

It is often convenient to publish an overall, relatively undetailed schedule covering the entire project effort. Then you can supplement this schedule with decisions and assignments made in project meetings for the next short period of time. Thus the overall schedule plus the last project meeting memorandum will be a specific guide for each engineer. In this way you avoid frequent changes to planned schedules.

There are many ways in which a set of tasks like the ones listed can be put into a schedule. The exact method you use is not important. But it is very important that a schedule be made and maintained. Furthermore, you will publish it in some kind of easy-to-understand form so that each person concerned can see what is expected of him and when his contribution is due. Many companies have their own forms and favorite methods.

For example, we might make up the schedule like Fig. 4-1. (You, the project engineer, working with your manager, have decided that

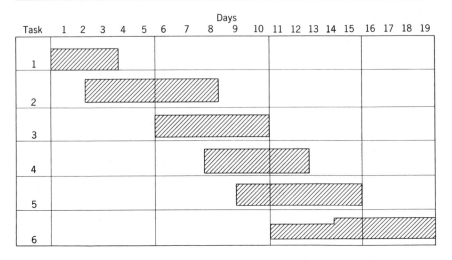

TASK · COMPLETION MILESTONE

1. Establish initial overall concept.   Distributed to project.
2. Determine critical factors in   List of critical factors for each as-
   each area.   signed area in hands of project en-
   gineer.

3. Specify all interface conditions   Interface boundary specifications
   and alternatives.   agreed to by each engineer involved and reviewed in project
   meeting.

4. Establish estimates for attain-   Characteristics reviewed in project
   able recorder characteristics.   meeting and accepted by all.
5. Establish final configuration and   Outline sketch of entire device com-
   make recommendation.   pleted. Major components specified.
   Memorandum for chief engineer
   prepared.

6. Complete proposal document   Proposal in mail.
   and costing.

Fig. 4-1. Project time schedules can be made in many different forms.

42

nineteen days are available and that it is practicable to complete the work in that time.)

It is of the first importance that each engineer know *exactly* what *he* is responsible for, so that he can see how his work fits into the schedule. It is usually not practical to put names on the schedule, so they are published elsewhere, probably in the memorandum of the first project meeting. Many of these assignments tend to change rapidly anyway.

The schedule of Fig. 4-1 shows overlap in the tasks. It is also desirable for you to indicate some specific *milestone event* as the end of each task. Tasks are often listed in the margin of the schedule under three columns: task number, task, completion milestone. Days listed in the schedule turn into dates when the job starts. A breakdown into five-day units suggests the flexibility of using Saturdays or weekends to catch up if overtime should become necessary. On a longer project schedules are usually made in terms of weeks or, less often, months.

Figure 4-2 is a cost schedule (budget) for the work. If this is not planned in advance, costs will always add up to more than expected. Such a schedule also becomes a clear meeting of minds between a project

Funds Required for Space Memory Proposal Work:

I. Project less proposal
   Time: Jones
         Kelley
         Bailey   100 hours each
         McKinley   @ 150% overhead   **$12,000**
         Smith (P.E.)   120 hours   **4,700**
      Support:
         computer time   10 hours   **1,000**
         Consulting   **3,000**
                                              $20,700

II. Proposal preparation
    (Details when decision is made on kind of book
    to be prepared.)

    Estimate, assuming book like that for Idaho
    computer job (includes 40 hours of P.E.'s time
    plus a total of 75 hours additional engineering
    time, plus usual reproduction group costs)   **$7,000**

Fig. 4-2. Cost schedules are important in project control and planning. They are not delayed until all details can be accurately estimated.

engineer and his manager on what the effort will need in the way of support.

The first time and money schedules are never delayed for complete, detailed data. This mistake has gotten many a project off to a bad start. You amplify and revise schedules as further information becomes available. In any event they will change as the project progresses.

Monitoring

Once good schedules are established they must be monitored. This does not mean that they can be carried out to the last letter as planned. Because engineering must be creative and must plow new ground, your project operation is never that smooth.

Monitoring means that *progress of the project must be continually measured against the established schedule.* It is a common mistake for poorly informed groups of engineers to work out good schedules at the beginning of a project and then never use them again. If the boss wants a schedule, they'll give him an eye-catching one to put on his office wall. Then that chore is done.

Actually the scheduling concept is a powerful *project tool* when properly used. It may be helpful to the manager, but it is ten times more useful to the engineers on the project. Your project group continually monitors its progress against the original plan as recorded in schedules. There are various ways of doing this.

Each engineer will keep the schedule before him in connection with his own work. Effective monitoring, however, usually requires a planned procedure. If left to chance or good intention, every contingency on the project will interfere with it. A good reason to postpone these checks of progress versus schedule always pops up. So, if you're the project engineer, have a routine, whatever it may be. On big projects of some length, perhaps you will want to monitor progress every Friday afternoon at two o'clock.

You can usefully combine these monitoring efforts with the regular (primary technical) project meetings. We have noted that administrative control and technical work are inseparable anyway. Here in the project meeting is where scheduling and monitoring become a powerful tool.

Most engineering jobs are full of unexpected turns and surprises. Some part of the work where real trouble was expected may turn out to go smoothly because one of the team members contributed an easy solution to a potential problem. Time and money originally scheduled there are no longer needed. Other areas, thought to be routine at first, may develop difficult situations which will require more time and money than originally planned. Without a sound, well-monitored schedule you will not notice these critical developments in time. But with the plan before you, group members can sit down together and work out the best way to use a new development or meet an unexpected problem.

In summary, you can't tell what is taking place on an engineering project without time and money schedules. It follows that these schedules are of little value unless the project is systematically monitored against them.

Controlling

First the project team makes realistic time and money schedules, showing its plan for conducting the work. Then it monitors actual progress against that forecast in the schedules.

This brings us to the last of the three divisions of administrative project control—those actions taken to change plans and efforts, to increase the effort in some areas of the project, to cut it back in others.

The controlling function adjusts work when needed so that overall time and money schedules can be maintained or at least be reasonably optimized. Having made a schedule, having followed it carefully to see where you are going off, it is then necessary to take appropriate action. This is controlling. You would be silly to just keep score (through monitoring) while the project slowly died technically or financially or for any other reason.

The whole purpose of scheduling and monitoring activity is to allow for actual control—for useful change in plans and work, in the best direction, and at the appropriate time, as needed. Don't be an engineer who mistakes scheduling and monitoring for control.

Practical controlling action is, of course, up to the project engineer. The big boss will sometimes be in on it, too. But each member of your project may have something to contribute at this critical point. In our illustrative case, who knows better than our chief mechanical engineer what changes might be made in the tape-handling mechanism analysis to help make up for a disastrous delay in establishing environmental specifications? Or who is better qualified than one of the electrical engineers to know that certain circuitry is turning out to be reasonably straightforward and to suggest that it be omitted from further work?

When a project group (systematically monitoring schedules at meetings) uncovers serious difficulties, it is natural for every member to join in working out a solution. The project engineer decides on some effective controlling action. Team members then support enthusiastically the decisions made.

Under these circumstances it is rare that problems cannot be worked out rather easily. Unfortunately, in some teams monitoring is poorly done or is left entirely to the project engineer or some outside group. Then solutions are much more difficult to find and carry out and usually are not timely.

The most common error in project control is not taking action soon enough. Poor monitoring action often causes this fault. Frequently, however, delay is due to a plain refusal to face facts.

As an engineer you are using the facts of nature to get things done, and so you eagerly hunt for them. Facts never work against engineering accomplishment but always for it. (They may be against your erroneous preconceptions.)

Although as an engineer you do not give up easily on your schemes, you can't afford to fight facts or feel hostile to them. And you certainly can't afford to ignore them, since they provide the solutions that you seek.

When it becomes fairly clear to you that a certain line of investigation (or a certain design approach) is not going to work out, stop it at once and redirect the effort. Delay wastes both time and money. Not only is a better approach not being worked on, but also time and money which would support a more desirable effort are being used up. If you delay the decision long enough, limited funds and time remaining will require a more drastic and unsatisfactory redirection.

Sometimes an inept project engineer or manager compounds this sin. He delays a needed controlling action. Later he takes measures which would have been appropriate (if started soon enough) but are now useless.

A second way to demonstrate poor project control is somewhat similar. It appears on the surface as foolish optimism but is really a little more complex. It begins when parts of the work start to fall behind. Engineers insist on "Just a little more time," and a vicious cycle is under way. As this trouble progresses it becomes clear to the outside observer that the whole project is in danger of badly missing its time and money schedules. Yet the project team insists that it is going to come out on schedule and will work harder to catch up.

It is important for you to see that "working harder" is not an effective control measure. Engineers work hard all the time, or they should. The more successful and challenging a project, the harder the team will work if well led.

Conversely, the poorer a project is going (and the more unrealistic the control), the harder it is for engineers to do their best. They waste energy in worry and in fighting discouragement.

This is not to say that planning some specific overtime work is not frequently useful if done for a clear purpose. Resolving that everyone will "try a little harder," however, is silly as a solution to a real difficulty.

A control measure (taken to meet difficulties) must be some specific change or augmentation which can be expected to solve the problem. Like "working harder," simply adding more money or time or people is seldom successful. Examples of good, specific control measures are the following: securing the services of an expert in the problem area for a specified time, abandoning one line of approach and starting another, giving up the area all together and doing the project in some other way, stopping further experimental work until a computer-supported analysis can determine some of the fundamental limitations, delaying further analysis until an experiment can determine whether what appears to be a critically limiting fact is valid.

Now, there is a not-too-subtle error in engineers' thinking which leads them to resist effective control. When an engineer is given a certain part of the work to do, he tends to identify it as his and to feel that its success is his success. The longer and harder he works on it, the more it is "his baby." He sees his original solution concept as the only one possible. If he has not been encouraged—even forced, if necessary— by his project engineer to participate in the project as a whole, he attaches himself ever closer to one phase of the work and one solution.

In this mental condition it is surprisingly easy to overlook or "wish away" the facts and constraints of nature. Unfortunately you then lose the very outlook which should provide you with a resolution of your problems.

As an engineer you can never go wrong by seeking out all the important facts which bear on your work, looking them squarely in the face, and then taking the action they dictate. To be able to do this, you must identify your success with that of the whole project, not with some limited part of it.

Richard S. Kirby and his co-authors pay tribute to some engineering of 1800 or 2000 years ago as follows:

"Roman engineers set their profession firmly upon economic principles. They employed exact specifications and detailed contracts. . . . Their public buildings, aqueducts, bridges, and roads show a sense of economy and efficiency just as the legal system of Rome reveals orderliness and analytical power. As Vitruvius expressed it, the engineers knew how to make a 'thrifty and wise control of expense in the works.' "*

---

* *Ibid.*, p. 59.

PERT Example

A powerful concept of control, including very systematic scheduling and monitoring—even computer-aided on large projects—has been worked out over the last few years. Its most interesting form for us is called PERT (program evaluation and reporting technique). We will briefly look at it here as an excellent specific illustration of the controlling measures discussed above, and a useful technique in itself.

First let us review briefly to bring out another idea: engineering projects vary widely in money involved, time needed, and numbers of people. In essence a project is a group of interrelated activities which can add up to provide for some specific want. We have seen that it is guided and controlled in terms of specifications, time, and money.

Now the organization and procedures discussed in Chapter 3 (and so far in this chapter) pertain to projects small enough to be done by *one team*—typically two to six engineers. The project team may be a little larger in some cases, and it could have supporting personnel like draftsmen, technicians, secretarial workers, consultants, and others. We use the term "project engineer" to mean the leader of such a team. Most of the rest of the book will apply explicitly to teams of this kind, although the book is generally applicable to the entire range of project situations.

You will find, however, that many projects are larger than one team can handle. For example, some space projects involve hundreds of professional people. (The term "project engineer" is sometimes used to designate the manager who controls one of these large projects, but that use will be avoided here.) The manager of a large project divides the work into successively smaller efforts. The smallest groups are project teams like those we have been discussing. Each is led by a project engineer responsible for that phase of the work. Integration between different project teams is completely analogous to integration between individual members of any one team. (We will consider in Chapter 12 some special techniques for accomplishing this higher-level integration.)

The PERT method of control was designed for very large projects. However, it has proven useful not only as a technique but also as a practical training device in control philosophy. For this reason PERT (or some similar scheduling method) is desirable on almost every engineering project. PERT methods are designed primarily for one-time or very probabilistic jobs. They should not be confused with another class of control techniques (such as "line-of-balance") which are used for repetitive or more reasonably predictable work, particularly in manufacturing.

The fundamental idea of PERT is to show graphically sequence and dependence relations of the various parts or "activities" of a project. To develop this idea and its usefulness a new project situation will be discussed.

(Whether or not you have used the PERT technique before, you need to go through the following example carefully. Work out every step and diagram on paper. Remember that your purpose in doing this is not to learn PERT, although this will be a fringe benefit for some. There are easier and more complete explanations of the entire PERT process available.* The real purpose here is to use this very simple example to clarify some of the control ideas we have discussed in the first part of the chapter.)

You have set up a small engineering research and development business near an Air Force base to meet flexibly and quickly Air Force needs on small contracts.

You bid on and receive the following job. The Air Force has the problem of procuring and depot-testing abroad a certain new type of crossed field tube. You are to design, build, and furnish a testing stand to be used with standard Air Force microwave test equipment, complete with instruction book, so that their technicians can run acceptance tests on the tubes. On your order they will furnish you one of the tubes as a sample. You will have to order the magnet from their vendor in this country. Since the first lot of production tubes is already on order, they are in a hurry and will airlift the equipment to an overseas depot when it is ready to ship. When you know approximately how large and heavy your equipment will be, a Colonel Jones at an Air Force base in Maine will give you shipping instructions, including packing requirements. At present he is on special assignment in the Arctic and will not return until eight weeks after your job starts.

Your contract calls for shipment eleven weeks after receiving authorization to start. You plan to provide the final report and instruction book a few weeks after shipping the test stand. A friend of yours who has a small machine shop business will take a subcontract from you for actual fabrication of the stand. He has an ample stock of everything needed except the tube and magnet. You already have in your laboratory

---

* For example, Joseph J. Moder and Cecil R. Phillips, "Project Management with CPM and PERT," Van Nostrand, New York, 1970.

a borrowed set of standard microwave test equipment for developmental work.

> *Requirement No. 1.* (a) Make up a PERT chart depicting the project graphically. What is the critical path to shipment?
> (b) What is the earliest and the latest time at which each of the events can probably be accomplished?
> (c) Can the shipment date be met as the job is now planned?

In PERT scheduling, as in any other scheduling, the first requirement is to "plan" the job, to divide up the work into logical parts, activities which, taken together, will accomplish the entire goal. It is necessary also to know the sequence of these activities and the way in which they must depend on each other—that is, which ones have to be done first before others can be started, or at least before they can be completed.

In the PERT method, however, scheduling and monitoring are not done directly in terms of activities, but rather in terms of "events." These events correspond to the "milestones" we used earlier in the chapter. They are specifically describable points in project progress.

Our project situation proceeds as follows.

After considerable analysis you come up with the tabulation of project events and interdependence relations shown in Fig. 4-3.

Thus the project work has been broken into activities described by the thirteen events listed; the length of time for each activity is estimated, and interdependence between activities has been analyzed. We chart the events as shown in Fig. 4-4. This chart is often called a "network." The estimated length of time to move to any event from all those preceding it is indicated (here in weeks) on each of the lines connecting events. Time moves approximately from left to right on the chart, so that events to the right occur later than those to the left (in complicated charts showing hundreds of events linked together, this may not always be feasible, and arrows on any line can be used to show dependence direction instead).

Time estimates used on the lines of the network are most advantageously obtained from the man or group that is to accomplish the particular activity and milestone. More realistic estimates can be had in that way, and at the same time the people submitting them commit themselves. They are involving themselves in the project control, as is necessary if the project is actually to be managed efficiently.

Symbol	Event	What It Follows	Number of Weeks to Reach after "What It Follows" Is Accomplished
A	Tube and magnet ordered	Start	1
B	Tube received	A	1
C	Magnet received	A	5
D	Test set-up design completed	Start	2
E	Shop work complete	D, B	2
F	Preliminary bench check on tube complete	B, C	1
G	Set-up ready to go for experiment and test	E, F	1
H	Shipping requirements requested from Air Force	D	1
I	Experimental and test data complete	G	2
J	Shipping requirements received	H	6
K	Equipment shipped	J, I	1
L	Report mailed	I, M	1
M	Instruction book written	I	2

Fig. 4-3. In PERT analysis the first task is to break the job down into significant events the accomplishment of which will ensure success for the whole project. The dependency relation between events is carefully considered.

Although only one time estimate is shown on each of the lines in Fig. 4-4, it is customary to actually obtain and record three estimates for each activity: the most likely time (used here), the most pessimistic time, and the most optimistic time. It is easier in this way for a good "most likely" estimate to be made, for everyone then clearly recognizes the chance elements in the work. Engineers who would be unwilling to quote a reasonable time estimate singly (since it may well be held against them) are able to be quite realistic on the three-estimate basis.*

For small and medium-sized projects it is often useful to carry the three estimates right on the network, as in Fig. 4-5. Those controlling a project are particularly interested in any real possibility of either a breakthrough or a difficulty, which could very materially shorten or lengthen the time required for a particular activity.

How long will it take from project start to shipment of the equipment? Figure 4-4 shows five different paths from start to shipment event. Elapsed weeks along each path must be added and compared. The

* It is also possible to use the estimates statistically to inquire into the probability distribution of time required to complete the whole project. This is usually worthwhile only on large projects. Programs have been worked out to make these calculations for almost every standard computer system.

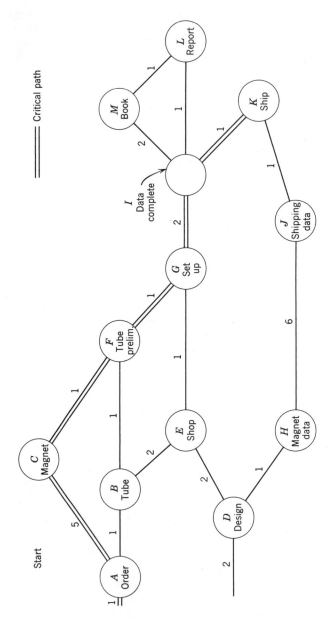

Fig. 4-4. PERT network for problem of Fig. 4-3, showing critical path to shipping. Numbers are estimated "most likely" time requirements in weeks. PERT networks depict clearly the essential relations between various project tasks.

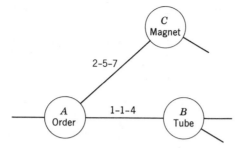

Fig. 4-5. Part of PERT network, showing triple times. Carrying the "pessimistic" and "optimistic" times on the chart helps to plan controlling action for more effective project administration.

longest path, timewise, is called the *critical path* and determines how long the project will take—*if* everything goes as planned, an important if! The critical path is emphasized by a heavier line in this chart and adds up to eleven weeks. Thus our equipment should be shipped on schedule.

Needless to say, it is the critical path which the project personnel will watch especially carefully. For if any event on the critical path is either late or early, the final date of the entire project can be affected. If, for example, the magnet were not received within five weeks of ordering, the whole project would be delayed. If, on the other hand, data gathering (item *I*) could be completed in one week instead of two, the whole project could be finished ahead of schedule.

It is useful to compute for each event the latest date it can be accomplished and still meet the schedule, and the earliest date it can be accomplished for a given start. Taking event *B* as an example, it can be accomplished no earlier than the end of the second week. But calculating backwards from the required shipping time, the end of the eleventh week after start, through path *KIGEB*. event *B* could be delayed to as late as the end of the fifth week. Thus there are several weeks of *slack* in that particular part of a path. Note that the latest date to accomplish event *B* could have been erroneously calculated by moving backward along path *KIGFB* as the end of the sixth week. The rule is to calculate all paths from the end event (in this case shipment) and take the "earliest late date." Similarly, in calculating the earliest date an event can be accomplished, as, for example, event *F*, where there is more than one path from the start, one should take the "latest early date" of all the possible paths.

Figure 4-6 tabulates these earliest and latest dates. Their difference

Event	Earliest Date	Latest Date
A	1	1
B	2	5
C	6	6
D	2	3
E	4	7
F	7	7
G	8	8
H	3	4
I	10	10
J	9	10
K	11	11
L	13	— — —
M	12	— — —

Fig. 4-6. A tabulation of the *earliest* date by which each of the events in the project example could be completed and the *latest* date on which it can be completed and still meet the scheduled shipping date. (End of week indicated.) "Slack" time for each event is difference of the two columns.

is *slack time* and shows the project team where opportunities lie to adjust the schedule without deferring the shipment date. In this example there is no slack in the critical path, that is, the project has been scheduled to take the entire time available. Although this is undesirable, it will frequently be the case. These dates and the slack times are often written in on the network.

The PERT network is sometimes used like a schedule, although it does not really constitute one in view of the slack shown. An auxiliary schedule, possibly contained in the minutes of the project meetings or supplemented by them, is usually required. In any event it is most desirable that the entire project team participate in the PERT scheduling and have the network available to them as the project progresses.

With the network and slack times established, the project engineer and his team are now in a position to easily monitor and control the project. Our situation continues.

Requirement No. 2. A week after the job has started you discover in the fine print of the contract (you should have read it more carefully!) a clear requirement that the instruction book be delivered concurrently with the equipment.
(a) Can you now make delivery on schedule?
(b) If you wish to expedite just one event to recover your schedule, which three events might be considered for this purpose?

From the way the job is planned, as shown on the PERT network (Fig. 4-4), it is clear that the instruction book cannot be ready within eleven weeks, and so the schedule will not be met. How simple this monitoring action is with the PERT data available!

Our second question (Requirement No. 2b) will be seen to be a control question. What should be done about this problem? How can the schedule be recovered? Actually the project engineer would probably use a combination of control measures to recover it. Here we will artificially limit ourselves to one measure to illustrate the possibilities of the PERT system.

The book will be one week late as now scheduled, so we must make up a week. The new critical path now runs from event *I* to event *M*. We must take this week out of the critical path. Since we cannot hope to recover a week from any event which is now scheduled for no longer than a week, only events *C*, *I*, and *M* can be considered for expediting, if we are indeed to recover the week from a single item. Event *C* would appear the most likely candidate if we can get the magnet vendor to agree.

Whatever control measure is taken to recover the schedule, it must be both specific and realistic. Let us assume that we decide we can reduce the time required for event *M* from two weeks to one week by providing for another technical editor and scheduling two nights of overtime work for the engineers involved.

Requirement No. 3. Two weeks after the job has started your Air Force customer tells you that it is urgent to change the delivery schedule to have the equipment and instruction book within the next seven weeks. He says that, if it will be any help to you, you can delay the final report another 60 days after the equipment is delivered. You promise to call him back shortly with an answer.

You call the magnet vendor and find that for an extra $500 he will strip a magnet out of his own test equipment so that you can have it in two more weeks. You tell him you'll call him back on this. You add your material overhead and other additional costs of the change, as well as profit, and find that the extra cost to the Air Force will be $750.

You call your Air Force customer and tell him what?

It is interesting here to note that, just because you can strip two weeks out of the critical path on the PERT network by expediting event *C* as described, you cannot necessarily meet the customer's request. Consideration of the network will show that with the time requirement for *C* reduced from 5 to 3 weeks the critical path changes completely to *DHJK*. Thus in order to meet the request some element of this path will also have to be expedited. Since the major time required is waiting for the customer to provide the shipping data, it would probably be best to inform him that for an additional $750 his request can be met if he can send the shipping data to you no later than six weeks from now.

Requirement No. 4. As a result of your call your customer decided to speed up the job as he had wanted to, even if it cost him a little more money and some trouble in his own organization.

Three weeks have now passed since the job started. You are shocked to discover in conversation with one of your technicians that the tube has not been received from the Air Force. (This is an example of poor monitoring.)

You first check quickly in your own organization to be sure that the order went out on time. It did. Then you call the local air base shipments officer. He informs you that the shipment was held up at an overseas base for higher priority cargo and will not be delivered for another week.

Can you now make delivery on schedule? If not, by how much will you miss? The solution will be left to you.

Control Support by Outside Groups

Control may be primarily administrative, but you can see that the controller needs an effective understanding of technical alternatives and project goals.

Some large engineering organizations have special groups which support the projects with scheduling service. Perhaps they make up and even monitor PERT charts or similar devices. These groups, which range from the very helpful to the worse than useless, are usually made up of senior clerical personnel who have little technical background (although there are exceptions). Pleasant and efficient girls are often chosen for this work.

As long as project personnel recognize that all phases of control are *their responsibility* and an essential part of *their own work*, the scheduling group may render valuable assistance. On a large project a good scheduler can relieve the project engineer of much detail in setting up and charting schedules and budgets.

The engineers themselves will have to make estimates and furnish information to be included in schedules. A scheduler with experience on varied projects can help the engineers to make better estimates and to avoid overlooking certain essential activities in planning. But the scheduler cannot make the estimate herself.

Schedulers put things in standard forms specified by management. Sometimes they can help a project engineer monitor his progress against the schedule and review it with management. On a large program (where computers are used to help schedule and monitor) schedulers will be familiar with the computer programs needed and can debug and run the problem.

On a small project it is more trouble to tell a scheduler the facts than to do all the scheduling work in the group, unless the engineers are learning how to schedule a project for the first time. Perhaps an experienced and trusted scheduler can review draft results and make suggestions. On any project, regardless of size, the use of scheduling support is a dangerous mistake if it induces a sense of relaxation and nonresponsibility in the engineers.

Sometimes scheduling groups are, or conduct themselves as, a Gestapo arm of management. This approach almost always fails. They cannot get reasonable estimates from the engineers and so resort to making their own poorly based judgments. This causes antagonism and consequent trouble.

One of the most successful scheduling groups that I have encountered, however, operated officially as a monitoring arm of top management. They supported projects with expert scheduling assistance, but they also monitored and reported on project progress with the cooperation of engineers, project leaders, and unit managers. On the infrequent occasions when a project leader literally refused to look at schedule and budget reality, the manager of the scheduling group actually called in top management to straighten things out!

The group's success under these conditions rested entirely on the character of the scheduling manager. He was a man of tact, great experience, and unusual judgment; he was respected and trusted by all in the plant, even when disagreed with. His group had cut its teeth on production problems before tackling the more difficult engineering design project schedules. Also, the nature of the company's business, with a great many projects running simultaneously—some in development, some in design, and some in production—favored this type of organization.

Control Your Projects by Looking Ahead, Not Back

Here is an example of another kind of control problem. In the central development laboratory of a large diversified company, one engineering section had several projects that involved information recording. One of them was a rather ambitious effort, really pushing the state of the art. It had been programmed for about two years and six hundred thousand dollars. I was called on to review it.

There were six months left and about half the money. The project engineer was quite proud of his "on schedule" performance. He showed me that his expenditure rate matched very closely the money schedule which had been set up more than eighteen months before. But he lacked a very good idea of how his technical performance to date matched his original plan. The plan was somewhat nebulous, undetailed, and hopelessly out of date. He finally managed to enumerate some of the important things his team had accomplished.

Of course, he and I immediately sat down to look ahead, to look at the *future technical steps still needed.* The project was drawing to a close. Certain items had to be designed and built. They were to have met specified tests six months from then. It turned out that things were not nearly as rosy as he had supposed, not really on schedule at all. The project team had been thinking almost exclusively in the past rather than in terms of the future.

Projects can be effectively controlled only in terms of the future—in terms of what remains to be done rather than what has already been accomplished. As you make the inevitable changes and adjustments during a long project, this point becomes more and more important.

A simple analogy will emphasize this idea. Suppose that you are flying from New York to Paris. The pilot, before taking off, determines the weather all the way, lays out his course, knows his headings and how he will change them on a time schedule. But even with extreme care and accuracy these plans will never get him onto the Paris runway. For this purpose he needs accuracies of a few feet vertically and a few yards horizontally. As he approaches Paris, he determines his position more or less continuously with respect to the goal—the airport and then the runway.

His New York planning is of little use now. If he stubbornly insists on sticking to it only an impossible piece of luck will prevent a crash. And what if the winds over the Atlantic change? His trip may be way off flight plan almost from the beginning. Does he look back to New York and his original plan to rectify this situation? Hardly at all. He is properly interested in where he is with respect to Paris and what he should do to get there.

So it is with the engineering project. Lay careful plans for the whole trip, but in adjusting and updating them to the project situation, look to the future, not to the past.

More effective project control is probably the single most needed improvement in engineering today. Both engineering group and individual engineer must improve. There are other major problems, too, but most of these are obscured by poor project control. Even management is sometimes unsophisticated here.

Good project control will contribute greatly in itself. It will also bring to light other professional engineering problems so that they can be recognized and solved.

Good Engineering Practices for Control

Scheduling

1. Schedule the whole project at the outset. Recognize interdependence of parts.
2. Identify the critical items early. Keep a current list of them.
3. Seek out the best time and cost estimates available on critical items.
4. Modify and update schedules as needed.
5. Bring all contributors or contributing groups into the sceduling process.

Monitoring

1. Follow and monitor performance (time, cost, technical progress) on a regular basis.
2. Include all contributors in this monitoring process, so that they are also "self-monitored."
3. Plan at least general alternatives for each principal contingency.
4. Keep the goal and its broad alternatives clearly in mind.

Controlling

1. Take early corrective action where needed.
2. Balance project effort on all needed phases.
3. Watch continually for places where effort can be reduced.
4. Make changes early rather than late.

Poor Engineering Practices for Control

Scheduling

1. Make no schedule or only a trivial schedule. Never go into the critical detail which will determine the success of the project.
2. Fail to recognize interdependence in schedule. Schedule unrealistically.
3. Make scedules in too much detail. Include noncritical detail. Include detail which is useless because highly sensitive to another contingency.
4. Select schedule milestones which are difficult to follow and assess.

Monitoring

1. Because of preoccupation with novel and challenging areas of the project, allow unmonitored tasks to run far off schedule.
2. Because of failure to identify critical items, do not follow these or provide alternatives.
3. Wait for other people or the turn of events. Raise no questions on schedule progress until critical deadlines have been missed.
4. Mistake proper rate of expenditure for adherence to technical schedule.
5. Allow an old schedule to become so outdated as to be useless.

Controlling

1. Mistake "scheduling" for "control." Fail to monitor or to take needed action.
2. Overemphasize certain areas to the detriment of overall cost and schedule.
3. Fail to make controlling changes in time.

Chapter Five

The End Product—
Drawings and Reports

If you are going to be a successful engineer, you'll have to be able to use language with ease—to express yourself fluently, confidently, and effectively. This holds for both written and oral presentation. You can't do either really well unless you can do both quite well.

Thinking in the context of Chapter 4—controlling projects in terms of the future, in terms of ultimate results—we come to a paradox. The immediate end of essentially all engineering work is paper.

"But how can that be?" you say. "Didn't somebody decide in the first chapter that engineering is not engineering unless it gets results—unless it goes all the way through to meeting some human need? What happened to that idea?"

Well, anything as complicated as engineering has to be worked out carefully in just about every detail. There are many people involved who have to communicate effectively with each other if they are to work together. You saw in Chapter 3 that *one engineer* can hardly take a job all the way through—even though he can never lose sight of the whole process—at least, in general. So each engineer (a specialist) has to pass on to the next one in the engineering cycle the results of his part of the work. (He must also communicate with himself, that is, store information for himself.)

In engineering, as in every other activity of modern civilization, paper is the medium whereby you do this. Our society lives on paper. We can't get along without it. (Maybe this is the simplest way to distinguish man from beast.) Engineering results are therefore put out as *pieces of paper.* This statement in no way contradicts the idea that these results must *then* carry through to fill human need.

So you see that the engineer's reliance on paper work is typical of all professional men. You are going to live or die on your paper work, and you must face that fact at the outset of any investigation into engineering excellence.

Don't Dodge Paper Work

Paper work is a tool to be used effectively, not a nuisance to be avoided as long as possible. It is indispensable to the engineer—a help, not a hindrance. Few engineers going out into practice seem really to understand this fact.

We saw in Chapter 2 why most engineers are biased in this area. You and I were attracted to engineering primarily through our interest in technology. Perhaps our interests were *limited* to such things. Thus, at the very time of life when we should have been learning to enjoy, appreciate, and use language, we were neglecting it. Although it is harder to cultivate verbal facility now, it has to be done. Every engineer has the best motivation in the world for making the effort—his success.

English is used for two rather distinct purposes—factual conveying of information, and esthetic conveying of emotional ideas. Naturally the engineer's use of language is mostly factual. But being a practical man he deals with people. They are his customers. They are his associates. They are usually his stumbling blocks. And people are emotional. They are unpredictable. Their decisions are never purely technical but on the contrary are almost completely or even entirely emotional. Emotional considerations usually sway people more than logical facts.

So you cannot neglect the poetic and esthetic aspects of English. They are in practice inextricably intertwined with factual use. Your reports must be readable. They cannot be boring or offensive. Your conversation must be convincing and believable. Would you wrap a costly gift in dirty newspaper? Then can you afford to present your excellent technical ideas (that might neatly solve your boss's engineering dilemma) in the "dirty newspaper" of poor or unimaginative or plebian English?

Engineering Drawings

The two principal forms of paper output for engineers are the drawing and the report. They have much in common, including a very similar purpose.

Most engineers have studied the intricacies of engineering drawing. They know how precisely and accurately and concisely it can define a material object—or, better, an engineering idea.

Paper tends to make honest men. If an idea won't work in practice, it usually won't lay out right on a drawing board. Thus the drawing process is more than a simple recording. It is in its own way an experimental test of the idea. Drawing won't take out all the bugs, but it will eliminate a lot of them.

Most companies have a very strict and detailed drawing system and regard it as sacred. Young engineers violate it at their own risk. Sooner or later, you run afoul of this system and learn to abide by it. These systems are all more or less alike in their broad outlines, but in each company it is the details that are important.

In general there are two types of drawings. Master drawings (often called *outline* or *assembly drawings*) picture the product or subassembly as a whole. Supplementary drawings (usually called *detail drawings*) define each part completely so that it can be procured or manufactured to fit into the overall assembly.

There are so many parts in most equipment that most organizations work out a complex method of numbering the parts so that you can readily find out which ones go into which product or assembly. Some companies do this with parts lists (sometimes considered a third type of "drawing"). Others relate parts to each other by a special system of numbering drawings.

Other special types of drawings are used in some places. Word specifications may be included in the drawing system, especially for purchased parts. Process data may be similarly included, in addition to the usual manufacturing instructions with regard to materials, finishes, etc., on parts drawings. Some organizations actually break the system down so that there is a separate drawing for each manufacturing operation.

Reasons for the fussiness of an organization with the details of its drawing system will become apparent in Chapter 9. Briefly the drawing system is used to ensure that the right parts are scheduled in the right quantity, both for the shop and from purchasing. Drawings ensure that the parts are compatible with each other and will perform a required function in the overall structure or assembly. If any drawings should be arbitrarily tampered with (even a change for improvement's sake) without going through procedures specified in system details, some mixup on factory floor or project site would almost certainly result. Parts would not fit together, or there would be an excess or scarcity of parts or some other problem.

One of the first things you do on a new job is to become completely familiar with the company's drawing system.

Draftsmen

An engineer has not completed his design work until it is correctly recorded in detail on drawings. Only then does he have reasonable confidence in its feasibility. Only then is the factory or some outside group able to manufacture it.

Needless to say, this whole operation of recording is *the engineer's responsibility*. You cannot delegate it to anyone else. You can get others to help you with it, but you alone are responsible that it is done and done right. It is a common mistake of young engineers to allow this responsibility to be diluted and their work blemished by not following a job carefully through to the conclusion of the drawing process.

The usual reason for this lapse from good engineering practice is a division of labor found in most engineering groups. Although engineers will often make up sketches of their work and design ideas, they usually give the actual drawing work to a draftsman. In fact senior layout draftsmen are called designers in most companies and have just that function. They take the engineers' rough sketches and work out details of placement (and often of materials) to make the parts fit together mechanically. In electrical work the draftsman will also, though to a lesser extent, trace out the wiring to be sure that it is consistent. If a draftsman is reliable, you can sometimes leave quite a bit of the routine design to him. This is excellent if you can do it.

But here is the point where trouble often comes in. An experienced designer knows a good deal more than a young engineer about the details of the work. Sometimes he isn't inclined to listen to that new kid in the engineering office. If you let him, he'll run away with the design, take it down some path different from what you want, and make your results poorer than they should be. Make good use of his knowledge; give him all the free rein and challenge that you can; but *guide the design work in the way you want it*.

You are responsible for the whole job. If there is going to be anything wrong with it, be sure that the mistake is yours rather than someone else's. The engineer will want to make frequent visits to the drafting room to observe and encourage and to work out problems with his draftsmen.

You will usually be asked to check and sign the drawings on your part of the job. There is nothing routine about approving drawings, even though your draftsman may tell you so. He'll say he checked them, and so he should. But if you are going to sign a drawing, check it carefully yourself. Be sure that if your name goes on it, it's right! As you come to have more confidence in a draftsman, your check may become a little less detailed, but check carefully always. Don't sign your name to a poor or incorrect drawing.

Good Drawings

Good drawings are distinguished from poor drawings in four characteristics. They are

> Complete.
> As simple as possible.
> Technically correct.
> Up-to-date.

A complete drawing is one which answers every necessary question about the part or process it describes. If the user has to come back for more information, the drawing is incomplete. What are the tolerances on this dimension? What are the material and the finish? Should we accumulate our tolerances from the left edge or the right edge? The drawing plus standard information in the drawing system should answer these questions and all others that will arise.

Good engineering is almost always relatively simple engineering. The whole engineering process is helped by clear, uncluttered drawings. Drawings are made to use. If they are clear and simple, they will be used with fewer mistakes and greater ease.

A technically correct drawing portrays a technically correct design. The part will work. There are no mechanical interferences. Electrical specifications are right. Every detail has been carefully considered.

In most organizations designs are under constant review and change, so that it is a real problem to keep drawings current. Drawings which do not reflect all changes up to the time they are dated will cause trouble later in production. When working with drawings always make sure that you are not using an obsolete one.

Automated Engineering Data Systems

With automation of manufacture and of data processing advancing rapidly today, some companies are eliminating or greatly reducing engineering drawings for manufacturing and design activities. Their place is taken by lists and programs which can be fed directly to or obtained from computers and data-processing machinery.

Although this is a radical change from the standpoint of production, it has little real effect on conception and design. Whatever means is used, something is required to *precisely communicate* the engineer's concept to whoever needs it, particularly to the production group. Similarly the production engineer needs specific information in some form on which to base his production process.

From this point of view at least, the form of the information medium is of little consequence.

Engineering Reports

The drawing presents carefully and accurately (and usably) the engineer's design work. His report presents with no less care and accuracy (and usability) his other work. The four characteristics of good drawings can also be used to distinguish good from poor reports.

Reports are written to be used. Although there are different types, in general a report is intended to cover the work of a certain project or part of a project. Thus the objective of a report is essentially the objective of the project.

Take, for example, a project which has been established to develop a new engine with certain characteristics and advantages. You aim the reports directly at these results. What characteristics are being found to be attainable or unattainable? What advantages over previous engines are revealing themselves? What characteristics cannot be obtained and for what reason?

The historical aspects of the report, while useful in most cases, are distinctly secondary to its real purpose. When a project is over, the final report is for most purposes the project itself. If $50,000 has been spent on the project, the management has in effect bought a report for $50,000. This is true even if a piece of developmental hardware has been produced. The hardware merely substantiates the results conveyed by the report and possibly served as a vehicle for arriving at some of these results.

In cases where a piece of usable hardware or a structure is produced, the report is, of course, only a part of the project. Even in these cases, the report stands for the entire project and should be written with this in mind. Suppose, for example, that the project is a large bridge. The final report (probably composed of a principal document and several ancillary ones) would outline what had been attained with regard to the specifications of the project. It would do this in sufficient detail so that the sponsor would have in his hand evidence of what he had bought for his several millions of dollars.

Let's consider our four criteria of excellence for reports.

Completeness in a report is its ability to answer every question about what was or was not accomplished on the project, particularly with respect to project goals. The report backs up its claims with evidence— usually experimental data or calculations—which enables the user to judge for himself the correctness of the results.

Like the drawing, a report should be as *simple as possible* in order to be most useful. Clarity and good organization are indispensable. If the report *is* the project or *stands as a symbol* of it, a confused, incomplete, difficult-to-read report will appear to reflect confused engineering work.

Technical correctness implies both that the report accurately reflects what was done on the project, and that the project work itself was well done.

An up-to-date report is timely. It is not uncommon on a poorly run project to find reports delayed extensively beyond the "end" of the project. This is really impossible since the project has not been completed until the reports are in. Reports are most often used as a background for decision making. When someone has spent thousands of dollars on a project to obtain information on which to base certain decisions, it is most appropriate that the project team provide him with a lucid and complete report on the specified day.

Report Writing and You

You will want to develop a good style and procedure for reports. Although larger companies usually help to develop their engineers along this line, ultimately each engineer is responsible for himself. You must go after this development aggressively.

Above all, *practice whenever you can*. Make each report you write— from the very beginning of your career—as good as you can. If you have time to rewrite it after it is finished, do so.

Examine carefully the reports that you read and use. What do you like about them? What is inconvenient? How could they be improved? What would make them easier to use?

If you will take these steps right at the beginning—or starting right now—you will not develop habits of distaste and consequent procrastination. Such attitudes toward report writing will blight your best technical work.

Like drawing, report writing tends to make honest engineers. If ideas are not good, they can hardly be made to write out well. Hence the report itself is a tool, also, to improve technical work. The clearer and more carefully done a report is, the more helpful it will be to the engineer who prepares it.

One hint will help most report writers: the excellent project team doesn't postpone report writing until report time. It gets material together as the project progresses. Careful entries are made in job notebooks. (This is discussed further in Chapter 7.) Intermediate results are written up carefully when they are determined, though probably not in final form, to obtain the technical consistency review that comes from writing.

When you are developing your writing ability through practice and review (and the use of books on the subject), keep in mind these five basic ingredients to good engineering report writing:

1. State conclusions specifically and clearly. These are the heart of your report.

2. Provide material to back up the conclusions and key it carefully to them. This enables the reader to judge for himself how valid the conclusions are.

3. Indicate all experimental procedures and conditions (and all analytical assumptions and methods) so carefully that someone unknown to the project team could repeat the work and get the same results.

4. Make the report quick and easy to use; for example, use
  (a) a good summary;
  (b) a table of contents;
  (c) good headings and articulate separation of parts;
  (d) pictures, drawings, and charts wherever helpful—key them carefully to the text and the text to them;
  (e) appendices for the bulk of data and backup material; ideally the main body of a report should be relatively short with references to more extensive material in appendices.

5. Write simply and directly.

On Large Projects

On large projects where reports may be extensive (perhaps several volumes) you can use various methods to employ more than one person for the writing.

The problem of using several writers for a report is similar to that of using a team of engineers on a project. It is hard to fit their work together into a complete and effective whole. But technical project integration has already been effectively done before the report is written (if it hasn't it's too late then). Hence the solution to integrating report writing is simpler and requires less effort on the individual team member's part. It is not necessary for each writer to follow the work of the others.

The key is to use one person to write the main body of the report. This is usually the project engineer himself. In most cases he should be able to do this with little input from his team, since for months he has been carefully leading the work anyway. The project engineer carefully outlines his report at the outset, sets the style, and estimates the lengths of various sections. He can then assign appendices and supporting write-ups to his team members, specify the data and material they are to furnish for the main body, and then draft the main body at one sitting. His team as a whole reviews his draft, or individuals can review and make suggestions in the areas covering their own work.

General progress reports can be handled in much the same way. Special reports dealing with a particular technical phase of the project will be most logically written by the engineer in charge of that part of the work.

On large projects the engineering team will sometimes be provided with an engineering editor or technical writer, particularly for a long final report. The concept of editing is usually more appropriate here than writing, although a capable person can rewrite, smooth, and organize material very helpfully. In particular he can save a great deal of the project engineer's time.

The writer or editor can't do much for individual team members. It is invariably more trouble for them to tell him what to write than to write it themselves (unless they are essentially illiterate).

Technical writers vary, like all other personnel, from the quite useless, to the valuable. The degree to which they are helpful depends, among other things, on how much they know about the general technical area in which the work is being done. A writer's usefulness is also very sensitive to his orientation on the particular project. If he is brought in at the last minute before the final report is due, he can contribute little other than simple editing, collating, organizing the material for reproduction, and carrying out liaison with the reproduction facility. Ideally he should be assigned (probably on a part-time basis) from the project's inception, attend some of the project meetings, keep abreast of progress and problems, and help plan the progress and final reports.

Even in this situation it is a very unusual man who can relieve the project engineer of the task of writing the main body of the reports.

How to Do It

One hears or reads frequent criticisms of engineering writing. Some are reasonable and factual; others, largely imaginary. The most common are as follows:

1. Engineers don't spell and punctuate well.
2. Their writing is stilted and pompous.
3. They don't write much, and what they do produce is hopelessly late and has taken too much time and effort.

Of course there is no such thing as an average engineer. Some excel at writing, whereas others have a great deal of difficulty with it. But the above points seem to cover the problem generally as I have encountered it in many years of practice.

The first criticism is of little importance. Secretaries and stenographers can be very reasonably hired. One of their duties is to take care of spelling and punctuation. Any engineer can quickly pick up the basic rules from the front of a desk dictionary.* It does seem that poor spelling and engineering talent sometimes go together. I once worked with a capable engineer who spelled even his name differently on various occasions.

The second and third criticisms, which go together, are serious. Unnatural and awkward writing is no less repulsive than Rube Goldberg's tortured engineering solutions. Good engineering is simple engineering. Good writing is terse and direct.

You learn to engineer by engineering, solving many diverse problems. You learn to write by writing. There is no other way: write, write, write! As an engineering tool writing is surely easier to learn than imaginative and effective use of the calculus. We forget the difficulty we first had with engineering technology. Watch a sophomore engineering student grappling with basic technical ideas. He has more trouble and wastes more time than most engineers experience with their first year of report-writing effort.

To summarize, here are two simple aids to becoming an excellent report writer:

---

* Although it is interesting to note that no less a man of action than Julius Caesar was fascinated with language and wrote a Latin grammar!

1. Practice engineering writing whenever you can. Never miss a chance.

2. Make everything you write as good as you can. Do this right from the start of your career.

Good Engineering Practices on Reports and Drawings

Reports

1. Plan the final report work as an integral part of the project. Start early.
2. Organize and write your reports for a predetermined purpose and audience.
3. Present material in terms of results.
4. Include sufficient detail (appendices) to permit a well-trained engineer to duplicate your work or to judge its validity and applicability to his situation.
5. Include a summary intelligible to someone quite removed from your project and field.
6. On all but the shortest reports, make clear topical divisions and include a short table of contents.
7. Make illustrations, graphs, and charts self-explanatory by means of captions.
8. Make illustrations, charts, and graphs attractive or don't use them.

Drawings

1. When planning the project, plan the drawing coverage in reasonable detail.
2. In planning the drawing coverage, consider specific objectives—model shop construction, production, government requirements, historical documentation, or others.
3. Meet with your draftsman on a daily or more frequent basis during drafting design to integrate the drawing work with the project.
4. See that drawings are complete enough to answer all the user's questions.

Poor Engineering Practices on Reports and Drawings

Reports

1. Regard reports as necessary evils to be written after the project is completed.

2. Write the report simply as a technical discussion of work done.
3. Write with the assumption that a reader already knows about your project and field.
4. Omit details which allow the reader to judge the validity and scope of the work.
5. Bury or so qualify your results and conclusions as to make any use of the work difficult.
6. Split responsibility for the main body of a report among several people.
7. Fail to tie illustrations and charts to the text.
8. Delay completion and publication of the report until many resultant actions from the project are already taken.
9. Fail to make clear, by reference or otherwise, what part of the work was accomplished on the project, and what part elsewhere.

Drawings

1. Allow drawing coverage of a project to grow haphazardly.
2. Make elaborate and detailed drawings when simpler drawings or sketches will do.
3. Fail to check drawings adequately. Leave the check to an unqualified person.
4. Fail to keep drawings up to date with respect to design or materials changes.
5. Fail to keep the draftsman abreast of changes so that a large part of his time is wasted.
6. Allow the draftsman to make fundamental design decisions for want of guidance.

Chapter Six

Problem Solving

An outstanding engineer, a man having eighty or ninety patents to his credit and continuing to be more productive than ever, told an engineering professor, "If you can just get one thing across to those seniors make it this: that there is no such thing as a closed engineering problem. All our problems are open ended. The engineer has to decide himself what he is going to do about them. No outsider can tell him, 'Do this,' or 'Do that.'" He went on to complain that some of his promising young engineers are often at a loss how to proceed with an engineering situation—what to do next. They apparently expect some superior to lay out the problem and tell them how to solve it. "If the boss could do that," he said, "it wouldn't be much of an engineering problem."

Although this man's group happened to be working on new developments, his comment applies to any really professional engineering work. Before those still close to their academic training get into actual practice, they often miss the whole point of problem solving. To intelligently practice engineering you must be able to recognize engineering problems when they arise. You must have a well-developed approach for solving them.

Engineering Problems versus School Problems

A major source of difficulty here comes from confusing the "problems" used in teaching engineering technology with real problems solved by engineers in practice. Poor problem-solving techniques are ingrained in engineering students as an accidental by-product of their technological training. It is hard for many to correct their approach even when they understand the difficulty.

Suppose that a class of freshman or sophomore engineering students is getting its first intensive experience with Kirchhoff's voltage law. The teacher puts a two-loop network on the board for a ten-minute quiz and asks for the power dissipated in resistor c. How will the students

73

solve this problem? By applying Kirchhoff's voltage law, of course. That's what the class is studying today!

The teacher expects his students to do this. He gave the problem because he knows that students learn by doing. He wants them to understand this law and its significance. He therefore makes them do something with it.

In engineering education this sort of experience goes on with little change for four of five years. To be sure, the work becomes more difficult and complex. But the idea is essentially the same. Here is a principle or a law or a method. Apply it so that you can understand and appreciate it. When you have done this well 2000 times, you will have a fairly adequate technological background to start an apprenticeship in engineering. The keener and more perceptive a student is, the quicker he will be to sense what kind of solution is desired of him—at what element of the course material the problem is aimed.

Actually, the solution method for such problems is quite straightforward. Apply the principle or method recently learned to the situation presented in the problem. This will produce a mathematical expression. Using the mathematical methods taught along with physical technology, solve the expression for whatever information is desired. Check your work carefully.

Now there is nothing fundamentally wrong with this method of teaching *technology*. And we have seen that the engineer must have a strong technological base. Unless he has modern technology at his fingertips, he can hardly practice engineering effectively. The method of teaching outlined above is effective and, most important, relatively economical in respect to time. But it does not teach *engineering*. In fact, as we saw above, it may tend to interfere with the real development of engineers.

In your own practice the problems that come to you are not aimed at any particular type of solution. You cannot relate them automatically to any course or set of principles. They present themselves simply as obstacles to your getting something done. You must turn them into opportunities to accomplish your goal.

What Is an Engineering Problem?

Let us distinguish more carefully between the two kinds of problems, those for the teaching of technology and those encountered by practicing engineers. We will use the term "example" for the former, and reserve the term "problem" for the latter.

"Engineering problem" is hard to define or describe generally, as are most fundamental human concepts. But here are some typical illustrations:

The efficiency of a newly built turbine generator is testing out 2% below design estimates.

A new electron-beam recording technique on which your group is working is not improving its effective bandwidth or its signal-to-noise ratio very fast. You begin to wonder whether you have overlooked some fundamental limitation.

Your boss directs you to visit customer X at once. He is said to be greatly dissatisfied with a new machine purchased from your company.

Maintenance requirements on a small subassembly of your product device limit design freedom greatly.

One of your friends remarks, "My company uses a number of special compressors with which we are dissatisfied because of the noise problem."

You read of a fairly new bridge being destroyed by a gale in another country. You are currently engaged in designing a somewhat similar one.

Your company has been manufacturing and selling a certain product quite successfully for five years with no change in design.

A customer with an unusually corrosive chemical bath desires some means of monitoring its concentration.

The production line making a product you have designed has just been shut down because of a high reject rate.

What a constrast there is between these problems and *examples* from technological teaching! In school there is little question concerning what is to be done. Solve this circuit. Find the deformation of that beam. Calculate the number of pounds of product that such-and-such a process will yield per hour under these conditions. But in the *problems* listed above there is almost no indication of what "technical" steps should be started for solution. In fact—and this is typical of many engineering situations—one is not quite sure whether there is any technical work to be done.

Maybe customer X need only be helped in reading the instruction book. Perhaps the bridge failed because of some blunder in calculation rather than a fundamental design problem. If that type of compressor cannot in fact be quieted, there is nothing to be done. No problem presents a business opportunity here for an alert designer.

Is there a problem, or isn't there? Apparently your first necessity is to investigate. You accumulate facts to decide precisely what the problem, if any, is.

In Chapter 1 we considered Thueson's concept of creative and responsive engineering, a differentiation based on whether the engineer acted primarily on the initiative of others or actually sought out and originated useful projects himself. This concept includes the handling of problems.

Does an engineer take the initiative to seek out and resolve engineering problems, or does he wait until his nose is pushed into them by the force of events, with consequent inefficiency and needless expense? For example, the fact that his company has been producing the same design for five years may be no problem at all. Change for its own sake is seldom desirable. On the other hand, investigation may show an opportunity to improve the product or process in the light of developments subsequent to its original design. The engineer who misses this chance may be brought up to date unpleasantly by a competitor who takes away his company's customers with a new design or a reduced price.

Thus *timeliness* is an important factor in handling problems. Intuitively it appears that there is a "right" time to undertake the solution of a problem, that is, to recognize that it exists. If action is taken before this time, little is accomplished. The technical state of the art may not be far enough along to support a new development. There may still be too much resistance to the idea within or outside the engineer's organization.

If action is taken after the right moment, much or all of the benefit may be lost. The state of the art may be changing again so as to render the new work obsolete. The market may be saturated with another incompatible product.

The question of *when* there is a problem is a matter of judgment, judgment developed by experience and through thoughtful observation of successful and unsuccessful ventures. It is characteristic of successful engineers that they are not often pulled about by circumstances but seem to create their own. They turn each successive problem into an opportunity through timeliness.

Now we can attempt to define or describe an *engineering problem* more generally as *a situation which will impede the accomplishment of some engineering goal or may provide opportunity for attaining an engineering goal.*

Problems start as an uneasy or an unhappy or a dissatisfied feeling on the part of the engineer. When you begin to have these feelings, you had better get busy and investigate.

How to Solve Engineering Problems

In one sense your work can be described as a continuous problem-solving activity. You organize and undertake a project to solve the problem that you think will be an opportunity. Work on such projects, in turn, is made up of a series of problems of slightly lesser scope, and these will break down still further. Finally we come to relatively simple and concise problems which appear superficially to be the examples used in training but really are not.

Each engineer develops a thoughtful and fairly uniform professional problem-solving approach. He becomes adept at coping with problems of all kinds. His approach can be conveniently broken down into six steps:

1. Define the problem.
2. List assumptions.
3. Consider available solution methods and select one or more.
4. Solve.
5. Check carefully, particularly the effect of assumptions on solution and vice versa.
6. Generalize and extend results.

The usual pedagogical example is limited to step 4, with possibly parts of steps 3 and 5.

Step 1 is probably the most important and most difficult of the six. In all but the simplest problems this defining statement is eventually written out. Step 1 includes initial investigation and a considerable amount of data gathering. During this effort the engineer's rather general (and possibly vague) definition of the problem becomes specific and concrete. An initial general statement of the problem is replaced by a detailed specific understanding of it. You decide exactly what problem, if any, to attack.

Step 2 is really a part of step 1; the assumptions are important in defining the problem. However, they are also listed separately for emphasis, since it can be fatal to think that a neat solution applies to situations beyond the original assumptions laid down.

For example, it might be assumed that if the price of a given product could be reduced by a particular amount the sales would increase in some specified way. However, the solution of an engineering economy problem on that basis could be very misleading if the assumption did not correspond with some new facts. As another example, in an electrical circuit problem you might make the assumption that distributed capacitances are negligible. But this will, of course, invalidate the solution for many high-frequency situations.

Steps 3 and 4 also go together, since in a difficult problem situation it will usually be necessary to experiment a little, to try several methods to see which is going to work out best. Now the young engineer is back on familiar ground with his technological training. However, not infrequently the new problem will at this point require a technique for solution which you have not learned yet. You recognize this, and go dig the novel or better method out for yourself.

Don't limit step 4 to analytical or mathematical solutions. Here the sky is the limit. In addition to these approaches you can go into the laboratory for solutions (this resource isn't limited to chemical engineers). Analog computers provide quick and valid solutions to many linear problems and increasingly to nonlinear ones. Then there's always the library; maybe you can find your solution in a book or magazine. But don't waste too much time among the shelves. In practice engineering problems of any magnitude are solved by a combination of methods.

Step 5 is a lot more than checking arithmetic. The entire problem definition and assumptions are examined in the light of solution results for consistency and reasonableness. Perhaps some mathematical or even laboratory experimentation is done to establish reasonable bounds on the solution. As an example, it may have been assumed that a slightly nonlinear phenomenon is substantially linear over the range of interest. The solution work may show whether this is indeed the case. Often simple experimentation, perhaps with an analog computer, will provide the answer.

Step 6, generalization, is making the most of the results, and indeed of the whole problem experience. Solutions to a specific problem may point the way to an entirely different approach to design. At the very least it is usually possible to generalize for a class of problems so that the same work does not have to be done over again. Computer programs developed for solution and "debugged" can often be generalized to accept a wider range of conditions. This is done most economically and effectively right after the solution has been reached, when the matter is still fresh in the minds of the engineers and their assistants.

Importance of Generalizing and Extending Results

It is at this point, step 6, that the unsophisticated engineer misses some of his best opportunities. We will see in Chapter 18 that the intense mental effort made to solve a hard problem tends to free mental processes to some extent from the limitations of habit. The presence of a new solution, in combination with this relaxed mental state, offers fertile ground for extended creativity. Instead of running back to bury yourself again in the immediate project, examine the consequences of the new solution in their broadest significance. The result may be new approaches to the problem area, whether it is a design, a production, or even an organizational problem.

A good example of the possibilities of generalization or extension comes from an investigation into dielectric amplifiers. After several years of work projects in this area had resulted in a good mathematical understanding of the behavior of resonant electrical circuits which included nonlinear dielectric material. Also, the particular material being used was well understood physically. But two mysterious problems defied solution: gain of the amplifiers was lower by a factor of two or three than theory predicted, and stability of the operating point was poor.

Finally one investigator, recognizing these problems and accumulating all the factual information on them that he could, hit upon the idea that there might be some significant temperature variation in the non-linear dielectric. Temperature had been supposed to be held constant by an oil bath. Heat calculation showed, however, that such variation did indeed occur, and resulting circuit calculations clearly explained the lack of gain and the instability quantitatively. Thus the problems were solved and checked well with experimental results.

The investigator could have gone back to his project with these excellent results and dropped the matter there, but he didn't. Examining the results of the solution broadly, he carefully considered what the mathematics were telling him. They showed that degenerative signal effects and a regenerative operating point effect were taking place simultaneously. From what he knew about resonance he could see that it should be possible to operate on another part of the dielectric's characteristic so that the signal effect would be regenerative and the effect on the operating point stabilizing. Experiment confirmed that this was feasible.

Thus the outcome of solving a problem was turned by generalization and extension from a mere explanation of why something would not work to a demonstration of how it could be made to work.

"The" Solution versus Iteration

One element of bad problem-solving habits sometimes unwittingly taught by engineering schools is the idea that each example has *one solution* and that it must be meticulously worked out in some rigidly *prescribed manner*. But any reasonably interesting engineering problem demands creativity. The work has probably never been done before or never done in a particular form or context. Thus in step 3 of our general approach to problem solving you will frequently devise an entirely new method, and you base this method on physical and mathematical principles which you suspect apply in this case.

Problem subject matter varies continually in practice. As you move from one area to another in carrying out your professional work, solutions will also be varied and new.

In applying the general six-step method, be flexible. There is much recycling here. For example, when step 4 is begun it may be found that a selected procedure will not work after all. But experience that you gain in the first abortive attempts will often point to a more promising approach. Similarly, solution work or ultimate checking may reveal the inadequacy of some assumptions or even of the basic problem statement.

Thus when you are in doubt about the original set-up of a problem (after a reasonable amount of thought) it is almost always best to proceed along some line even if it appears inadequate. In this way the right approach may be brought quickly to light.*

All experienced engineers habitually solve problems by an *iterative process* which begins with an estimate of the answer, continues with refinement by approximation methods, and concludes with the use of whatever precision techniques are required. In most cases the engineer has some sort of estimated answer before he even starts to calculate carefully, even though frequently this first answer is quite crude. The first estimate may come from experience with similar or related problems or often from an almost unconscious mental computing ability.

The first estimate is important because in most cases it gives a rough independent check before you are biased with a wrongly computed answer. It often provides an answer good enough so that you can discard an idea without further calculation. It becomes a part of your "judgment" so that you can quickly check the more detailed calculations of subordinates.

Two additional aspects of professional problem solving, from the many that could be discussed, will be mentioned briefly. You will find some others considered in later chapters.

It is surprising that so many young engineers neglect accuracy requirements in problem solving. Or perhaps it shouldn't be surprising since little attention is given to this need in the examples used in school

---

* Many helpful techniques in problem solving are presented throughout the book, "Conceptual Blockbusting," by James L. Adams, W. H. Freeman, San Francisco, 1974.

to teach principles and methods. It is necessary to carefully consider *both* the accuracies needed from solutions and the accuracies available from original problem data. It is useless to calculate with greater accuracy than is needed or than original data warrant. Once numerical expressions are put down with a specific number of significant places, they tend to become "accurate" to this number of places for most people who see them, including their originator.

The great loss of accuracy entailed in subtracting nearly equal numbers is another surprisingly common cause of trouble.

Are You a Problem Hobbyist or a Problem Solver?

Engineers love problems, and many become so absorbed in them that they waste needless hours in study and solution. The best defense against this is to keep the goal and application constantly in mind.

Regardless of how interesting the physical analysis or mathematical manipulations may be, effort and time expended must be kept proportional to usefulness and results. This requirement will often dictate methods or even decide whether certain problems are to be solved in any detail at all. Step 6, generalization, can be such a gold mine that it should never be neglected. But it can be overdone.

Because engineers have been (and still are!) graduated with a label of mechanical or civil or electrical or something else, there is a tendency for a man to think of himself as a body of specific problem-solving techniques—perhaps thermodynamic techniques—looking for certain kinds of problems to solve. This attitude greatly limits engineering accomplishment.

In his everyday work an engineer is *problem-oriented* rather than discipline-oriented (that is, rather than problem-solving-technique oriented). Finding a need that you believe you can help to meet through the application of technology (and probably working as a member of a project team), you uncover the problems to be solved in a timely way. Using an orderly mental process, you solve them with whatever methods are needed. A man uses what he knows, learns more, consults with others. When he has the immediate solution, he goes on to generalize creatively and to extend the new ideas he has uncovered.

Help from Computers

That large digital computers are producing a revolution in problem solving no one can doubt. Computations that were once too costly or time-consuming are made with ease. Some predict a drastic change in the mathmatical training of engineers. Formal analysis, they say, is no longer as important as it was before the day of the computer.

However that may be, every engineer will want to avail himself fully of this new tool. The only way to understand what a computer will really do is to *take a simple course in programming and put some problems on the machine.* Then your thinking can expand in every direction, finding ways to use computers in your work.

But the point that will interest you here is the beautiful example that these machines offer of what we have looked at earlier in this chapter. Now that you know what an engineering problem is and is not, it is apparent that a computer can't find a problem or solve one. It can be a big help in both activities, but it can't take the place of the engineer. Somebody has to tell the computer exactly what to do (program it). In most cases someone must interpret the results.

Strictly, all that a computer can do in our six-step engineering problem-solving approach is step 4—solve. Conceivably it might be used to help in other steps of a very complex problem. But it really is still helping with step 4 of the subproblems that they represent.

A computer has only two advantages, and many disadvantages, over you in step 4. First, once it has been told what to do, it can do this over and over again at fantastic speeds. Second, it almost never makes a mistake of its own, although it will help you to make a lot of ridiculous mistakes you would never make otherwise. This is one of its principal disadvantages, but remember that people who used to drive horses laughed at the first automobilists for the same reason. A horse had enough sense not to run into a tree or go into a ditch. Nevertheless there are many more automobiles than buggies around today.

No engineer would ever suppose that he could take a computer into a project meeting, ask it to ferret out any problems that needed attention, and have it take care of them, or send it down to Mobile to investigate a customer complaint. This is the lesson that the computer teaches—that step 4 by itself is not engineering problem solving. In many ways it is only the easiest part of the process. So how could anyone who aspires to practice engineering ever suppose that, even without a computer, step 4 techniques alone would constitute real engineering problem solving?

If an engineer is going to err in computer use, he should err on the side of using one too much rather than too little. Too many engineers have not yet bothered to learn what is happening to their profession (and maybe to their own employment if they are this far behind the times!).

But there is a reverse danger. A project group in which I was interested was falling behind. On investigation it turned out that the project had bogged down for a week because a computer was backlogged with work. It would be another week before the computer could spare the 15 or 20 minutes to make a necessary calculation. The problem involved some solid geometry in connection with a satellite. With a little encouragement the unit manager and his project engineer went to a blackboard and in an hour or so had an analytical solution to the problem in functional form. This was in lieu of a computer solution in the form of points on a curve, obtained a week later.

Be sure that your computer is your slave and not your master.

Good Engineering Problem-Solving Practices

1. Recognize that the biggest part of an engineering problem situation is to decide what the problem, if any, is.
2. Look for the problem or problems involved in any situation that is out of the ordinary or disturbing.
3. By good judgment discover and attack problems at the right time— not before a solution is possible or useful, and not too late to reap maximum benefits from the solution.
4. Attack problems with a systematic procedure.
5. Check problems to include a careful consideration of the effect of assumptions on the usefulness of the answer and the effect of the answer on the validity of the assumptions.
6. Solve problems by an iterative process which starts with an estimate and continues with successively more sophisticated solutions until the results required are obtained.

Poor Engineering Problem-Solving Practices

1. Confuse the scholatic examples used in teaching technology with real engineering problems, thus failing to recognize and attack a problem when it comes along.
2. Expect the boss to lay out the problems to be solved and to be satisfied with an academic answer.
3. Attack problems piecemeal, seeking only such information as is essential at the moment.
4. Neglect to make clear the assumptions under which the problem is solved, and to consider the effect of these assumptions in checking.

5. Fail to extend and generalize on the solution.
6. Assume that there is only one solution to a given engineering problem or one set method to attack it.
7. Use solution methods with far greater accuracy than is required or than is justified by the kind of data available.
8. Assume that an engineering career consists of looking for the types of problems that one has been trained to solve.

Chapter Seven

Laboratory Work and Experiment

One of the commonest ways of solving problems in engineering practice is by laboratory experiment. We will examine some of the reasons for this. Work in the laboratory (that is, with physical equipment) is usually an important part of your early experience, and it will continue to be important to you in one way or another as your career develops.

There are few indifferent engineers in the matter of laboratory work. Most engineers are either very fond of working with physical equipment or dislike it heartily. The first group would use the laboratory to solve just about every problem encountered. The second is convinced that analytical solutions are the only respectable parts of engineering problem solving.

Actually neither group is very near the truth. Laboratory work, like any other engineering procedure, is a means to an end. It is only one specific tool to be used for solving engineering problems. Like any other tool, it has its best applications and is to be employed with discrimination. You will want to examine this tool, determine its characteristics, learn what its best uses and limitations are, and see what you can do to keep it sharp and effective.

It is possible, though usually awkward, for a beginning engineer to avoid much laboratory work. Naturally some will enjoy laboratory work more than others and be more successful at it. Even though you may be one of the few who will succeed in your profession without real laboratory talent, you will want to understand this tool and be able to have others employ it for you when you direct the efforts of a group. For the majority of us, who will spend a substantial part of our time in a laboratory or with working equipment, little more needs to be said about why we are interested in this topic.

Laboratories in Industry Are Different

Let's start by comparing and contrasting school and industrial laboratories from an engineering viewpoint. The primary, almost exclusive, purpose of laboratories in an engineering school is *to teach*. In the laboratory the student learns by actual doing with equipment. He investigates important physical facts, and because he does this himself, they stick better and mean more to him than if classroom work or lectures only were employed.

Now in a hurried way you as a student picked up certain good laboratory habits, techniques of getting things done with physical equipment. You began to gain an understanding of and a patient persistence with the everyday difficultes that are always encountered in practical work. But you had time only to scratch the surface. From school laboratories you can become reasonably familiar with quite a bit of equipment—oscilloscopes, for instance—that you will use in practice. This know-how and these good laboratory attitudes, however, will have to be carried much farther in industrial work.

Similarities between school and industrial engineering laboratories end about here. The purpose of the industrial laboratory is not to teach (although you can't help but learn in one) but *to solve practical engineering problems* and to test engineering ideas, devices, and products. You will acquire much secondary information in the laboratory, some of it of great interest. A small part may even turn out (in spite of planning) to be of crucial interest to your project. But the purpose of going into the industrial laboratory is primarily to find a specific answer to a specific problem.

What Kind of Laboratory Problems?

Problems of Ignorance

One reason for solving some engineering problems in a laboratory is that we do not yet have a thorough enough grasp of physical science to always design by analytical methods.

Apparent discrepancies between theory and practice are often commented on disparagingly by many, usually with little understanding. As an engineer, you know that you live in an ordered and predictable universe. True theory coincides perfectly with practice. Cases which seem to belie this are merely examples of incorrect theory. Often theory is incomplete—it does not really comprehend the situation to which it is applied.

Fig. 7-1.

A clear, elementary example is the basic alternating-current electrical circuit. If we connect an a-c source to a resistor as in Fig. 7-1, we can calculate current flow simply by Ohm's law. If we measure the current flowing into the resistor carefully with an ammeter, theory and practical result coincide completely. A designer could confidently use Ohm's law as the theory to design these circuits.

But now suppose that there is some requirement to physically lengthen the conductors *a-a'* and *b-b'* of Fig. 7-1. As they become longer and longer, the correlation between theory and practice becomes increasingly poorer. When they become substantial parts of a wavelength at the frequency of the source, theory "breaks down" and is of no help at all.

What has happened? Is our theory incorrect? Surely Ohm's law is still valid. But as a theory it is inadequate or incomplete to cover what is going on in this new circuit. In a sense it is misapplied, although that is surely not the fault of the law. We need to add some more sophisticated concepts about transmission lines to explain what is happening. Persons unacquainted with transmission line theory might complain at this point about the unreliability of Ohm's law! Instead they must investigate further to see what *assumptions are inherent* in the original limited theory that no longer apply in the case at hand.

You can extend this simple example (although it wasn't simple at all to the engineers who first encountered it!) to situations of greater interest today. For instance, the strength of materials has never been understood theoretically. All valid predictions of material strength have always been based on laboratory testing. With the advances in solid state physics of the last few decades there now appears hope that an adequate understanding (theory) of strength can be developed, but this knowledge is not yet available.

Problems of Convenience

A second reason why you go into a laboratory is that you can solve some problems experimentally much more easily than by paper analysis, even when the theory is well known. For example, radio engineers want to know the antenna pattern from aircraft in flight or the best place to mount aircraft antennas. To work this out theoretically, even with large computers available, is usually tedious. The spatial relation of each few square inches of aircraft skin to antenna must be considered. This problem can be solved quickly in the laboratory by constructing accurate wooden models of the aircraft in question, coating them with a conducting paint to simulate the skin, mounting scaled antennas wherever desired, driving the antennas with small power sources whose wavelength has been scaled with the model, and taking field strength measurements at reasonable distances from the model.

Similar examples exist in the scaling of dams and river channels. There are many others in various fields.

Since no engineer can be abreast of all theory available for problem solving, it is sometimes easier to try an idea in a laboratory than to learn the theory involved. Of course, this approach has some severe limitations.

Problems of Checking—Insurance—Realism

Even when quite good desk solutions are available and are used for development ideas and design, an engineer is always concerned about whether or not he has really taken everything into consideration. When you have been working on an idea or a design for six months and the day of completion approaches, you will lie awake at night and wonder whether it is really going to work. Failure can be quite embarrassing. I have seen some red-faced engineers at crucial moments.

Therefore a third reason for going into the laboratory is to check out a concept or design that has been arrived at by other means. You want to see what problems, if any, remain. You want to establish confidence in the earlier work.

Often this kind of laboratory effort is useful to convince superiors and others who may not have had your analytical background. By demonstration you can convince them that an idea is a good one and that therefore they should continue to back it with funds and other support.

Testing

A fourth type of laboratory activity is testing. Completed products or components (or their prototypes) are usually tested in some way.

You operate them in a manner as close as possible to their intended service. Careful measurements are made to see how well they perform.

For example, if you are working for an electric power company and are buying a new steam turbine, you will be very particular to see what you are getting in the way of efficiency and overall performance. You observe the manufacturer's tests of the machine and conduct your own when it is installed. At the other end of the spectrum, even materials coming into a plant are checked and tested in some way to ensure that they meet the specifications to which they were purchased.

The Judicious Combination

Only a very limited number of worthwhile engineering problems can be solved by analysis alone or by experiment alone. Most are best solved by a *judicious combination of analysis and experiment*, although the same person may not do both. You can combine experiment and analysis in a great variety of ways, which can be illustrated by another, almost absurd example.

Suppose that in the circuit of Fig. 7-2 you want to find the ohmic value of heater resistance R which will produce exactly 500 watts of heat. Certainly you can work out this problem analytically. Could the laboratory enthusiast solve it experimentally? He could get a calorimeter to measure the heat generated in the resistor over a period of time and by substituting various values of resistance on a 100-volt source could find the right one for an output of 500 watts.

But there's another, easier method (adding some analysis). If our laboratory enthusiast knows that power into the resistor is the product of current and voltage, he could measure the current and voltage as various values of resistance are tried until the correct one is found. Or, better, he may be aware that power is a monotonic function of resistance in this circuit. He could speed up his experimental work greatly by bracketing the power and making straight-line interpolations

Fig. 7-2.

until he has correct resistance. No doubt you would conclude, however, that the judicious mixture for this problem is 100% analysis and no experiment!

A more realistic example could be the design of a mechanical tape transport system for a recorder. Here a beginning can be made by analytically considering the masses, permissible tensions, speed and acceleration requirements, drive characteristics, and so forth. It would certainly be essential, however, to get parts of the system into the laboratory early to observe performance and measure results. Friction is difficult to predict analytically. A continual interchange between analysis and experiment will be required to solve stability problems and to optimize component placement.

The design of a special-purpose electronic amplifier offers another good example of the usual combination of analysis and experiment. Designed first on paper, it will be breadboarded in the laboratory and the final component values "trimmed up" by experiment.

It is often dangerous to rely on either analysis or experiment alone. When you rely on experiment for design without any real understanding of the physical situation, it may be that what works in the laboratory will not work in actual field conditions. Normal component tolerances may upset things. Analysis will usually show how sensitive a solution is to variations under manufacture or in operating conditions.

On the other hand, to rely solely on analytical methods for problem solving courts disaster from overlooked elements of the problem. How can you really know that your analysis embraces all significant aspects of the physical situation?

Engineers concerned with large, costly, one-of-a-kind structures such as dams, bridges, or large missiles can hardly afford to experiment with whole systems. Nevertheless some means of gaining confidence in a design must be worked out. Often critical parts can be tested separately. Models are built. Critical elements in the new design may be found in use elsewhere and observed. A great deal can be done with computers to simulate actual testing.

Notebooks

If you are going to solve most problems by judiciously combining experiment and analysis, it is reasonable to inquire whether good laboratory work does not have many of the characteristics of good analysis. Since the purpose of laboratory work is problem solving, the general problem-solving approach discussed in Chapter 6 is helpful here.

Orderliness, system, clear approaches to what is to be done and what has been done are just as valuable in laboratory work as in calculations. We might suspect from Chapter 5 that paper is useful in the laboratory. The wise engineer keeps a laboratory notebook in more or less diary form with dates for each entry.

Start your project by writing down in your notebook what the problem is, in some detail. For each phase of experimental work state the object of the experiment; make a careful sketch of the set-up; record equipment numbers. Data and calculations follow. Neatness, thoroughness, and orderliness in your notebook save hours of time later and sometimes save results.

But most important is the fact that *a professional approach to notebook keeping in the laboratory is mentally catching.* It will make your thinking and laboratory set-ups more orderly and systematic, so that your results will be better.

Good *laboratory* notebooks have proven so helpful in project work that in many companies engineers now use *project* notebooks. These include not only the laboratory data but also essentially all thinking and work on the job. In other organizations this material is included in project memoranda written up frequently by each contributor and circulated among the team. Other groups have a weekly summary report from each team member. Whatever system is used, there seems to be no adequate substitute for a well-kept notebook in the laboratory itself.

Your Technician

Either at the outset of your work or soon thereafter you will be given a technician to help you, especially in connection with laboratory work. These people are invaluable when properly used. Learn how to work effectively with your technicians. They are members of the project team that we discussed earlier. Sometimes you have them on a continuing basis; sometimes you get one just for a limited phase of the work. Many of the conclusions that we developed about group work are applicable here.

If your technician is good, he will be better than you are at many phases of laboratory work, and you can learn from him. If he is a poor technician, you can help him grow and develop. As in your work with draftsmen, be sure that, if the job is yours, you run it. But give your technician all the responsibility that he can handle. Let him develop and apply his own ideas to the job in the areas where he is competent. Take him completely on board the project and keep him fully informed.

Here are two perfect examples of what the relations between an engineer and his technician should and should not be, respectively. One of my technicians, Bill, was an excellent constructor and experimenter with perhaps 15 years of electronics experience. He was a radio amateur, a gentleman, well educated. When he was assigned to engineer Henry, they worked together smoothly and efficiently. Bill was never at a loss for what to do. When Henry was out of town, Bill went right ahead on the job; he was familiar with just about every detail from vendor problems to the customer. If a crisis came up in Henry's absence, he solved it or knew when to bring it to me. While Henry was doing most of the analysis, handling the customer, and running the overall job, Bill did all the ordering and specifying of parts except for the most critical elements.

Bill handled about 90% of the shop contacts. He built most of the equipment in the laboratory as he and Henry designed it. Henry even let him design electronic elements like power supplies and some of the amplifiers. Frequently I found them together in the laboratory up to their elbows in electronic equipment. Just as often, I found Bill working there alone while Henry analyzed logic diagrams and circuits at his desk, with time out for frequent visits to the laboratory.

When that job was over I assigned Bill to a younger man named Ed, a very capable engineer incidentally. Two weeks later Bill came and complained to me. I had seen the trouble brewing but couldn't stop it. Ed told Bill almost nothing about the new job. He didn't really make much use of Bill's technical ability. In the laboratory I would often see Ed buried in the equipment while Bill just stood around, wondering exactly what he was doing, handing in tools to Ed, turning the lights off and on as needed. As good a worker as he was, it took only two weeks for him to get fed up with that kind of relationship.

On the other hand, it would be equally foolish to assume that you as the engineer will do the paper work and your technician the laboratory work. He isn't usually equipped to handle that responsibility. Laboratory work is such a vital part of a project that you must be involved in almost every detail of it. If you are not interested enough in equipment to want to operate it and experiment with it yourself, you had better go into some other line of work.

Not all of your technicians will be as good as Bill, but it's up to you to train and develop them. You will get out of them what you ask for. Technicians are no different from engineers in that respect. We all learn by doing. We all grow by having more responsibility thrust on us (within our ability to strain and handle it). Learn how to work effectively with your technicians and every other member of your team.

Some Good Laboratory Practices

Laboratory work is time-consuming, but some engineers seem to spend a great deal more time than others for the same results. Here are some of the reasons.

Laboratory instruments are often out of repair or adjustment. Experimenters instinctively feel that a new-looking electronic voltmeter or oscilloscope with a respected name plate can be trusted implicitly, but it is not unusual to find after several days of frantic effort that most of the trouble is coming from faulty instruments rather than from the equipment under development. Don't be fooled in this way! When you make a set-up, first run simple checks on all the instruments. This small investment of minutes will save hours later on.

Similarly, laboratory instruments are very complex these days. The engineer coming into the laboratory with some vexing or fascinating (depending on which kind he is!) problem to solve wants to wade right in and get at the work. But much time is saved by first learning what the instruments will do and how to use them. Get the instruction book. This is a good example of the old adage about making haste slowly.

Much time is lost in laboratories by being sloppy with set-ups. A careful, neat set-up takes a little longer, but it almost always pays dividends in time and money later. In addition, it's desirable to have an orderly work place when the boss drops by to see what you're doing.

Laboratories often become very cluttered and unsightly. Piles of unused equipment accumulate on the benches, on shelves, on the floor under the benches. I have seen laboratories where the engineers spent two days off and on looking for a small piece of equipment that was buried under the clutter, trying to make do without it, returning to the search, getting mediocre results with a substitute, and so forth. This is nothing short of disgusting. Keep order and make your technicians observe good housekeeping in the laboratory. Have a weekly or more frequent white-glove inspection. Remember that disorder is catching, expecially among technicians.

In some instances time is lost through trying for too much perfection in set-up. A good rule is to have it in perfect, uncluttered order but as simple as possible to begin with. If you start big and complex, you will lose time trying to get everything right, and many details will be changed before long anyway. Most laboratory set-ups are temporary at best and should be so designed. How much effort to put into a set-up is a matter of judgment. While it's better to err on the side of neatness and careful construction, you shouldn't err very far, especially at the beginning.

Some engineers become so interested in side effects that they waste days before getting back to the work for which they entered the laboratory. Keep a sense of direction. What is your mission? Which hill are you trying to capture? By all means keep your eyes open for peculiar effects as the scientists do. In rare instances they may be the key to your work. But don't be dragged far down side alleys just as a matter of curiosity, scientific or otherwise.

The final time waster is a little more subtle. You can waste days and weeks by carrying on experiments too long. Here the engineer has forgotten his laboratory mission completely or has never defined it exactly enough to know what he is really after.

In designing experiments make a definite decision as to where you will stop or at least how you will make the decision to stop. Like an engineering project, the experiment seldom stops by itself. It is an old saw among engineering managers that they have to personally stop every experiment. Decide when you have obtained about all out of the work that you are going to get. Then stop the work yourself.

The Biggest Time-waster

The biggest obstacle to crisp, effective, engineering performance in the laboratory (or when working with equipment) is failure to *define the problem precisely* and failure to keep it defined.

In poorly run engineering groups people are always going into the laboratory to "try an idea" or "see where the problem is." These objectives are in general commendable, but their generality reveals the naivety of the speaker. Never do anything with equipment without first making a careful statement in your notebook of what you are trying to do—what you are after.

No doubt this first statement will be inadequate. You will find changes to make, redirections that are needed. By all means make them, thoughtfully and again precisely. When you become confused or discouraged, stop and write up a summary of what you've learned so far and where you'll go next. Laboratory work without corresponding thought (and this usually means writing) will quickly degenerate into futile knob-twisting.

It is common experience for engineering managers to find a discouraged member of their group, worn out by several days of futile laboratory work, ready to give up his part of the project as impracticable. The young engineer, on being required to write up his results as they are, is surprised and encouraged to discover how much he has really accomplished. The problem is clear again. The direction in which to go next is often obvious.

Good Engineering Practices in Laboratory Work

1. Define the problem clearly and set down the specific end sought before starting laboratory work.
2. Design the sequence of experimental work carefully to move as directly as possible toward what you're after.
3. Plan the work in terms of equipment and facilities needed, time schedule, laboratory assistance.
4. Check test equipment first.
5. Alternate periods of experimental work with write-up, analysis, and thought on the overall problem.
6. Arrange to have much of the detailed construction and data taking done by a laboratory assistant. The engineer doing the work keeps completely conversant with every significant detail.
7. Record every element of the experimental work in such detail as to enable someone else to repeat the experiment with equipment having the same serial numbers.
8. Plan and execute deliberate minor variations in experimental conditions to provide understanding of stability and repeatibility under practical conditions.
9. Insist on a high standard of order and cleanliness in the laboratory. Maintain safety standards consistently.

Poor Engineering Practices in Laboratory Work

1. Go into the laboratory to putter with "a problem."
2. Plan to think out results after laboratory work is complete. Spend all your laboratory time "trying" things.
3. After wasting hours looking for "bugs" in the equipment under investigation, find them in the test instruments.
4. Incur several hours' or days' delay because of unforeseen equipment needs.
5. Allow the laboratory to become so cluttered that time is lost in hunting for things.

6. Make such an elaborate and polished experimental set-up that little real work gets done.

7. Make such a hasty and sloppy experimental set-up that most of your time and effort is wasted chasing set-up trouble. Damage equipment by consequent accidents.

8. Put off write-up and all analysis until laboratory work is ended, thus being unprepared to intelligently discuss work with those interested, forgetting significant details, omitting important elements of experimental work.

Chapter Eight

Design

Design is the heart of the engineering process—its most characteristic activity. If you and I are going to understand engineering, we'll have to understand design.

Think back to the five-step engineering process—conceive, experiment, *design,* build, test. Design is right in the middle of it. Sooner or later, if you're really going to get something done, you will have to come down to earth and do it. Design is that point. Up to then you've conceived and experimented. You have lots of new ideas—some look pretty good, and some don't. But now it's time to make up your mind, to decide exactly what you're going to do. That's design.

If you like definitions we could use that one. *Design is deciding exactly what you are going to do.*

In the more or less formal engineering design process there's another aspect that needs heavy stress—documentation. That's tied to the word "exactly." If you haven't gotten your decision down on paper—in the form of accurate and complete drawings—then it's fair for the boss to say, "He hasn't really decided yet. He hasn't finished the design." Hence in engineering you will habitually think of design as deciding and getting your decision down on paper.

The following steps—build and test—are the proof of your design pudding. Building is usually the most expensive part of engineering. Can your design be built economically and with the facilities your production organization has? And when it is tested, will the thing really work? Will the bridge or building meet the original requirements placed on it? Will the machine or whatever else you have designed operate properly?

Although you may not be engaged directly in making designs on paper for production, your engineering work is always tied closely to this design idea. It's useful to think of most other engineering activities as being organized to support design.

The industrial researcher, for example, works to uncover new facts useful for design, even though he himself may not be so directly moti-

97

vated. The development engineer is working to bring these new ideas to such a state of feasibility that the designer can confidently include them. Market researchers uncover facts that help the designer to know what customers need and want from his design. The engineering manager is concerned to so run his total organization that his engineering designers receive the guidance and close support that they need to produce excellent, useful, marketable designs.

Now, in addition to designing the product itself, which may be a table radio, a huge dam, or an improved automobile, there are other types of design. If we look again at the five-step list, it suggests ideas like designing experiments, designing test equipment and test procedures, working out manufacturing methods. The manager designs effective organizations.

This concept of deciding on something new to meet some requirement certainly pervades all human activity and is not a private prerogative of the engineer. In his own work the engineer will find himself designing continually, even though it may not be in direct relation to a product configuration. Much of the philosophy of this chapter will be quite broadly applicable and useful for any thinking engineer.

Don't be fooled because we talk here primarily in terms of product. That is simply a means of focusing attention on design concepts.

"Best" Design—"Optimum" Design

One of the most significant yet usually unacknowledged facts about design is this: *there is no formula or procedure to arrive at an optimum design.* It is important for an engineer doing design work to realize this truth. There is no known way to find the best design for a particular situation or requirement.

Because of the way in which engineering education works it is easy to assume that engineering problems have ideal or best solutions. But look at any (but the most simple) design problem. There are literally endless ways in which the design can be varied.

Take an automobile engine, for example. The placement of parts with respect to each other, the exact shape of the castings, the precise size of parts—all are susceptible to substantially infinite variation. And we have not considered the more obvious choices open to the designer—number of cylinders, stroke, bore, materials, lubrication system, ignition arrangement, number of carburetors, head clearance, and others.

Or consider a radio receiver circuit. Even if the designer limits himself to standard values of commercial components, there are thousands of combinations which will work at least fairly well. When he goes on to consider the placement of parts on a chassis, the possibilities again become substantially infinite.

Thus the design problem in engineering is certainly not to consider all possibilities and find the best one. This would be impossible. Probably for most significant engineering problems today there is no best solution anyway. It is easy to see how frustrated a design engineer feels to think he is being asked to find (by some magic) *the* solution. The words "best" and "optimum," frequently used in design discussions, amount almost to synonyms for "good" design or short cuts for "the best design that can reasonably be had in a given length of time."*

Time and Design

Francis K. McCune, Vice President for Engineering of the General Electric Company, has pointed out the unusual importance of time in design engineering work. He notes that the price obtainable for a new product drops rapidly for several years after the need for it becomes apparent. This occurs because various competing devices and designs quickly enter the market to serve this need and to reap the rewards it makes possible.

Thus speed is essential in design, and you can properly trade off design refinement for time. In effect, Mr. McCune points out, "If the profit margin were low and the price decline rapid, it wouldn't take very much delay in a design project to turn a profitable venture into a loss."†

---

* The term "design optimization" has been popular for several years. Scholars from various fields are reading papers on the subject. Many are worthwhile; some are, at the least, deceptive. These papers are of two general types.

The first recognizes the problem that we have stated here and attempts to restate it symbolically and to draw some conclusions from manipulation. In this sense we are indeed engaged in optimization—in trying to do the best engineering we can within the constraints of time and available ideas. But it is not simple mathematical optimization in the usual sense, for the inclusion or exclusion of new ideas is usually of far greater importance than the quantitative handling of known inputs for the design.

The other type of paper on this subject deals quite literally with a conceptually straightforward optimization of results from a limited field of known variables. Even this work is of interest and use to the designer as long as he realizes that it is helping him to solve only *a part of his problem* and hardly touches the vitally important element of recognizing the possibility of new ideas and untried approaches.

† Francis K. McCune, "Presentation to Second Conference on Education for Business," July 31, 1963.

After a product is no longer new and its price begins to stabilize, it often becomes worthwhile to considerably refine an original design.

Your Design Problem

If you are assigned to create some design, recognize at the outset that you will not literally optimize. Your problem is to choose or configure a sound, practical, reliable design from an unlimited number of possibilities. There are always many more designs that are poor or won't work than there are good ones. Usually, however, a large number of good designs that will work acceptably are possible, so many in fact that it is impossible to consider them all individually. Don't try to.

It follows that there is substantially no design, no matter how much effort and talent have been lavished upon it, that cannot be improved. Hence it is very difficult to stop a design project. There is always an alluring reason for the engineers to continue, to make obviously available improvements. So, again, no project will stop itself. You have to call a halt.

Designing for Function

A second important principle that simplifies your approach to a design problem is this: solution is sought in terms of performing a *function* rather than designing a part for its own sake. A cam in a machine, for example, is there to accomplish some function. It is designed in terms of the function. A transistor amplifier similarly is designed in terms of the specific function it is to perform rather than simply as an amplifier. This kind of thinking frees the designer to recognize radically new solutions to his problems, often including the elimination of a part or the combination of parts.

Design Constraints

One common failing in design solutions, which causes a great deal of trouble, is what can be called a *peaked* design. A narrow or peaked design is one in which the effects of normal variations in manufacturing or use have not been adequately considered.

Some radio receivers provide good examples of this error. They are designed for use on a 117-volt a-c power source, but the designer who does not also consider that a nominal 117-volt house supply can easily vary by plus or minus 4 or 5 volts is heading for a poor product. We have all seen home radios whose tuning is affected when an electric dryer is switched on or off. Unless the house wiring is inadequate, this is simply an indication that the radio designer has a solution which is inadequate off the peak. He peaked his design for 117 volts and did not adequately consider the requirement that his product be operative in an environment which could include power supply variations from 112 to 122 volts.

There are many other examples. In mechanical design many wonderful things can be accomplished by specifying very close tolerances on parts. When you realize, however, that an ordinary shop cannot consistently hold these tolerances, you will avoid that kind of peaking also. You must consider the environment in which your design will be manufactured.

We can usefully generalize from this idea through the concept of a *constraint*. A constraint is something which constrains—that is, which forces, compels, or obliges, or which confines forcibly. The range of supply voltages the radio designer must allow for, or the manufacturing tolerance limitations the mechanical designer must take into consideration, are constraints on their designs.

It is useful to think of design in this light: the design engineer is generally faced with a requirement to accomplish some function within a number of constraints.

Living with Constraints

The idea of operating effectively within a number of constraints is not a new one. In society, for example, you are expected to make a living within some very definite constraints, including abiding by the law, not harming others, and paying taxes. A merchant operates his business freely but within constraints of the market place: his customers' tastes, the financial health of the communty, his competitors' activities, the weather, the availability of goods, to mention a few. The clothes we wear can be varied greatly, but there are limits of custom and fashion and availability.

Unfortunately the word "constraint" is usually taken with a somewhat negative, restrictive, and unpleasant connotation. This must be why engineering designers sometimes overlook these important factors. Actually, the constraints on design, like social or economic constraints, are better considered as *guides, directives,* and *opportunities.* They are part of the overall environment in which you work.

The successful person, in life or engineering design, usually has as one of his major characteristics the ability to work along smoothly, not fretting about constraints or fighting the problem. He cheerfully and alertly accepts his constraints, partly as rules of the game to be used to greatest advantage, and partly as challenges to be overcome with ingenuity and creativeness.

Important constraints that stand in the way of obtaining objectives—of making a design adequate or significantly improved—are called *limiting factors*. Limiting factors which you can overcome and turn into break-throughs to better design are called *strategic factors*.

We will also continue to use the term *critical factors* to designate especially important elements of a project. They may be those elements that can be turned into strategic factors through good engineering. Or they may be limiting factors of such severity as to prevent doing the project at all.

As a capable engineering designer don't waste time and energy in wishing away constraints or trying for an optimum solution. Take all constraints into consideration as a matter of course. You will creatively discover one or two so-called constraints which can be shown to be no longer valid (that is, limiting factors which can be transformed into strategic factors). By exploiting this new situation, you will come up with a superior design.

Here is an example taken from the design of television receivers. For a particular table model, marketing requirements might be listed as follows (these will be seen to be some constraints on design):

> Sell for less than $200.
> Have at least a 20-inch tube—preferably larger.
> Weigh less than 20 pounds.
> Look as flat as possible.
> Be as small as possible dimensionally.
> Receive fringe signals well.

Here the experienced engineer might come up with a design that met every requirement except the second, questioning whether for the normal function of this type of set in a modern home so large a screen is really useful. Thus the second requirement might become a strategic factor. Several years ago a home appliance company actually went through this thinking, recognized that the screen size race among com-petitors had about reached its reasonable limit, quietly brought out an excellent small-screen receiver, and immediately increased its share of the television receiver market by a large percentage.

A more technical example of a strategic factor might be the following case. Because of high current requirement an electric bus structure is very heavy, necessitating expensive structural supports and massive insulators. The heavy current requirement and resulting copper bus weight constitute severe constraints on the design. An alert engineer uncovers a newly developed, light-weight conductor material, thereby eliminating a severely restrictive constraint. Here the bus material became a strategic factor.

As another example, explosion-proof equipment* had been designed for many years to depend on rigid (and therefore heavy) enclosing boxes with wide flanges. This constraining factor was later overcome for low-power signal circuits by limiting the energy available to go into an accidental spark. Here the concept of designing for overall function was most usefully exploited.

A less alert designer might have limited himself to recognizing that the purpose of the enclosure was to prevent a gas explosion within it from propagating to the outside, as indeed it was. But some engineer transcended this viewpoint to realize that the *function required* was simply to prevent an outside explosion. Therefore preventing an ignition in the first place could be an adequate and economical solution, eliminating the heavy box entirely.

You will certainly be limited if you think always of how things were done in the past.

Where Do Design Ideas Come From?

New engineers worry sometimes about how well they can design. Some think that designers have accumulated a vast fund of their own information upon which they draw for each new design. The beginner doesn't know where or how he will get his own ideas.

It's true that an experienced designer does acquire a lot of information through practice. Also, he naturally has a certain confidence and sureness derived from previous successes.

But the sources of design information are surprisingly informal. Anyone can use them. As long as no significant constraints are missed, a fresh new idea is of greater importance than detailed design data from the past.

There is no substitute for experience, but experience comes very fast when you design alertly, considering (but not limiting yourself to) what has been done in the past. You will want to think about what is being

---

* Equipment designed for safe use in explosive atmospheres, as in gassy coal mines or on aircraft carriers.

done by competitors, what new developments bear on the problem, and the whys of these things.

With few exceptions a man who has designed a certain kind of device once knows almost as much about it as the man who has done this ten times. The difference in their subsequent performance will lie entirely in the new ideas that each can bring to bear.

The most important sources of design ideas are the following:

1. Fundamentals of technology that the engineer knows or can glean from textbooks.
2. The old design and other related designs.
3. Periodical literature.
4. Other people.
5. The engineer's own creativity.

The Old Design

The old design or a related one is an excellent source if properly used. It would be ideal to find an adequate design and use that rather than going to the trouble of starting all over. But if you have been set the task (by your boss) of designing a specific item, probably there is something inadequate about the designs available. Look at them carefully and see why they are unsatisfactory. Your creativity can then suggest a way around the inadequacies, thus producing the elements of the new design. Of course you will want to think from a *functional* standpoint and to avoid being trapped into the same shortcomings as the previous designer. The original design is a working and adequate design for its own purpose, however, and so can carry a good deal of weight in your thinking.

Unfortunately there is a major pitfall in using the old design this way: it tends to stifle creativity. Once an engineer sees one way to accomplish his goal, it becomes much harder for him to detach himself and think out a completely different method.

A good habit to cultivate in solving any problem—design or other—is to take your own careful look at the situation first. Do this before you bias your thinking with anyone else's opinions or ideas or with what was previously done. After a few hours of thinking on your own you can more safely look at the past. In this way you are less likely to miss a good but radically different approach. Don't let the old way or ancient prejudices bias your thinking when you first start.

It is equally foolish, however, not to consider all sources of ideas, including other people and old methods. A common fallacy among young engineers (and some older ones) goes something like this: "The boss gave me this design job to do. I am an engineer. I must be creative. Therefore I'll go back here into my corner and invent a new way to do this. I won't pay any attention to what my colleagues think, especially the experienced old hands. The boss wants my ideas. I have to come up with something new all on my very own."

Of course this is ridiculous. The boss probably doesn't want this man's ideas at all. He wants a design—a good design or, even better, an outstanding design. But he doesn't care where the ideas come from.

This fallacy seems particularly troublesome to new project engineers. They think they have to prove themselves, and they do. But the proof of good project engineering is good design. It does not necessarily have to be entirely original, so use all the good ideas you can get hold of. Give honest credit where credit is due. But don't be afraid to use the best ideas available, and don't be caught using any other kind!

Very often the best design is a simple modification of the old one. Don't be afraid of this. There are a few so-called engineers, however, who never offer anything else. The excellent engineer will want to look at other alternatives and be sure that a simple modification is really best.

Chance Sources

The creative designer finds a strategic factor on which to operate. In practice, this almost always involves applying some new idea to the design. For example, a new weldable aluminum material is developed. The designer can now, for the first time, apply it to parts in an airplane passenger compartment. Creative design takes advantage of this change in the state of the art. The material becomes a strategic factor. The previous weight constraint of the old material and fabrication methods is overcome.

As another example, an engineer designing radar circuitry may, while browsing through a design magazine, come across a new capacitor mounting for printed wiring boards that some vendor is offering. Perhaps this product solves one of his current vibration problems.

Now there are a substantially infinite number of design possibilities, and the designer is always looking for applicable new ideas. Hence chance appears to enter into finding strategic solutions. You therefore put yourself in positions where you will be likely to run across ideas. Read several design magazines in your field regularly. Scan ads. Go to conventions and examine competitors' and other equipment carefully. Listen to papers at meetings. Eat lunch with other designers. Be awake, alive.

Thus the design process is a fascinating combination of logical processes and chance, of hard work, judgment, and inspiration. The man who tries to limit it to textbook logic and so-called optimum solutions will be a worried, frustrated engineer.

Design Efficiency

Critical Factors

In a military operation the commander weights the main attack. To do this, troops elsewhere are thinned out as much as practical. These are principles of war called "mass" and "economy of force."

Similarly you *put major design effort on limiting factors* and turn one or more into a strategic factor. To do this, efforts on the more straightforward portions of the design are minimized. In design work it is useful to list the constraints. Select the limiting factors and analyze each briefly. Choose those which will first be pushed for strategic factors. During design, periodically review this list and analysis. Development of the design and gathering of information remove some items from the critical list and add others.

The big danger in design work, as in any other problem solving, is that some important factor may be overlooked. There are two possible penalties: (1) in spite of effort and excellent results in some areas, the equipment or structure as a whole will not function satisfactorily, or (2) some much better solution may be missed.

Poor project groups solve the easy and more or less straightforward problems first. In the meantime several large critical problems, which may torpedo the whole effort, are unrecognized or put off on one pretext or another. To prevent this you must keep a careful, frequently consulted, up-to-date list of what your limiting factors are. Put almost all your effort into these items.

Standard Items

One essential means of economizing, in terms of effort, time, and costs, is using standard components and materials. Most large engineering organizations insist on this and have special groups to publish, monitor, and enforce standards. These save so much time and expense in design and manufacturing that they are used even when good design must be distorted slightly to accommodate them.

For example, a company may have a certain series of capacitors from one or two vendors which is its standard for that type. They have been tested and used for some time, and smooth procurement arrangements are established with good prices. The designer needing an intermediate size not included in the series, which could be procured from another source, will in almost all cases conform his design to a capacitor from the standard series. This saves his own time, too, in most cases. He has all the data he needs on the standard line at his fingertips; there is no need to wait and worry while a qualification test establishes whether a new part actually comes up to the claims of its manufacturer.

As another example a designer can, where feasible, take many elements from an older design in noncritical areas, even when he could make small improvements by redesigning. For a new truck he might use the old rear end design from an earlier model, if that were not considered a limiting factor in his new design goals. Design economies like these provide the time and funds needed to make a really strong effort in critical areas.

Because of the tendency of poorly managed engineers to violate standards needlessly and continually reinvent what is already available, standards groups sometimes become quite deaf and unreasonable about the need for a new part or procedure. Where a new nonstandard approach can become a significant strategic factor, the designer will make vigorous efforts to see that it is accepted. He is in a much better position to do this if he really understands and appreciates the need for standards and for the work of these groups and has cooperated effectively with them in the past.

Eliminating Alternatives

An engineer starts with an almost unlimited number of solution possibilities and narrows them down quickly to as few as possible. He is not greatly concerned lest he eliminate through haste one idea that might be *slightly* better. He is concerned that he not eliminate through haste a *decidedly* different approach which could be a *great deal better* than the ones he has retained. Frequent recycling in the light of further experience on the project will help here. Select the several best general approaches which suggest themselves and analyze them in detail. Critical alternatives result. You can save more time if you examine these alternatives in parallel as far as possible.

For example, a not uncommon type of alternative is the following. Accuracy requires that certain components stay within a small percentage of their nominal ratings for a given application, but the design must be good over a wide temperature range. The designer has resolved this critical area down to two alternatives: (a) procure larger and considerably more expensive components which are supposed to be less sensitive to temperature, or (b) use standard components and temperature-control the whole compartment with a thermostat and heater. He is not sure which is better or quite how either should be done. Both alternatives should be pursued in parallel instead of changing to one after the other has failed to give adequate results. The engineer starts his laboratory technician testing high-quality components in the contemplated circuit. At the same time he finds vendors for the heating and heat-control equipment, obtains telephone specifications on it, and can quickly get started on a test drawing of the possibilities there.

Sometimes, when there is an extended delay in resolving some critical alternatives, a project can be speeded by designing the noncritical parts so that they can accept either of two critical possibilities or can do so with simple modification. By this kind of parallel aggressive action, *pursuing the design information that you need instead of waiting for it to come to you,* you will break through more quickly and economically into an outstanding design.

Use analysis, laboratory work, and confirmed vendor information to solve your problem, and to establish the confidence and sureness you and your superiors need that the design will be functional and adequate.

Nonoperational Constraints

A number of years ago a successful manufacturer of television sets came out with a new line with advanced design features. From a purely operational standpoint these sets were the best or among the best in the industry and available at competitive prices. But the designers overlooked a major constraint: the chassis were assembled in such a way that it was difficult to get at them for servicing. For a television repairman to make either minor or major adjustments and replacements he had to use a great deal more time and trouble than with other sets.

Although these sets were so good that they needed less service than many others, of course things did go wrong sooner or later and then the trouble started. Repairmen hated them and complained to the owners. Some even refused to service that particular line. Others added an extra charge to their flat rates. Those who charged by the hour warned customers of the bills they would run up. Busy servicemen put those sets at the ends of their lines so that owners were without them for days at a time.

Wherever television was discussed by servicemen or set owners, this particular company was run down. It did not recover its good name and business position for several years. In addition, the furor must have adversely affected some of the company's other appliance lines.

If the designers who made this mistake were thinking exclusively in terms of performance constraints, they must have been rudely awakened to the facts of engineering life. It is easy for most designers, being technologically trained, to take into consideration the technical performance requirements of their designs. For example, in a radio receiver one thinks immediately of sensitivity, hum level, selectivity, fidelity, power consumption.

To see the very real nonoperational constraints, however, you must come out of your technological shell. You must live in the real world where you attempt to better meet human need. Examine every aspect of how your product will interact with that world. Think from its conception and manufacture through to its final demise and disposal. It is in the areas beyond technical performance, important as this is, that a great deal of design work falls down.*

We have already observed that complexity is generally undesirable. Complexity makes the design itself less sure since there are more things to be overlooked or misunderstood. Good designers will scornfully say, "Anyone can design anything if you let him make it complex enough." Simplicity marks the excellent design, although often you can't avoid some degree of complexity.

Beyond the design stage, producibility is a first major constraint. You will have to continually keep in mind how each part is to be made. What are the specific abilities and limitations of your manufacturing facility? If you are not familiar with them, become so at once.

Reliability in use is another major constraint. Will the product continue to work? What is the cost to the user of a failure? Would it be better to give up some degree of performance in order to ensure a longer reliable life? What design techniques will influence reliability?†

The product's environment sets important constraints in some cases. Something may work fine on land. Will it perform as well after several months of exposure to salt spray? Will the heat dissipation of your device be adequate if it is mounted near an internal combustion engine?

Installation requirements are sometimes overlooked as a constraint.

---

* In an excellent book on engineering design, Morris Asimow says, "Engineering design almost always requires a synthesis of technical, human, and economic factors; and it requires the consideration of social, political, and other factors whenever they are relevant" (*Introduction to Design*, Prentice-Hall, Englewood Cliffs, N.J., 1962, p. 2).

† "The Pons Fabricius [at Rome] built in 62 B.C. is still standing" (Richard S. Kirby et al., *Engineering in History*, McGraw-Hill, New York, 1956, p. 70).

If a very stable gyroscopic device for maintaining direction is to be produced, how will it be aligned in the first place, or periodically thereafter?

The user or operator sets definite constraints on design. Does the equipment require more performance on his part than he is capable of? Can he really understand the instructions? Can he be depended on to perform the required maintenance? Is the customer willing to mount training programs for his employees? What kind of training will be needed?

Can the design be maintained and repaired satisfactorily? What changes would make maintenance easier? Could the product be designed so that at the end of life cost could be recovered by turning it in for rebuilding?

The major constraint in all design and problem solving of any kind is cost. Although most engineers recognize this as far as the product itself goes, many fail to consider the costs of nonoperational constraints. In considering costs for design purposes one must think in terms of total costs, including design, manufacture, test, training, installation, maintenance, and salvage.

For example, in modern buildings more costly materials are often used for certain purposes because maintenance requirements are thereby eliminated. Palleting is expensive to start with but can save handling costs later, and so forth.

George Washington's appreciation of these aspects of engineering design is shown by this letter written to his London agents in February, 1764:

"We have been curiously entertained of late with the description of an Engine lately constructed (I believe in Switzerland, and undergone some Improvements since in England) for taking up Trees by the Roots; among other things it is related that Trees of a considerable Diameter are forced up by this Engine, that Six hands working one of them will raise two or three hundred Trees in the space of a day; and that an Acre of Ground may be eased of the Trees and laid fit for Plowing in the same time. How far these assertions have been amply realized by repeated experiments it is impossible for me at this distance to determine, but if the Accounts are not greatly exaggerated such powerful assistance must be of great utility in many parts of this Wooden Country when it is impossible for our Force (and labourers are not to be hired here) between the finishing of one crop and preparation for another to clear Ground fast enough to afford the proper changes either in the Planting or Farming business.

"The chief purport of this letter therefore is to beg the favour of you Gentlemen to make minute inquiries into the Tryals that have been made by Order of the Society and if they have proved Satisfaction to send me one of these Engines by the first Ship bound to this River (Potomack). If they are made of different sizes, I should prefer one of a middle Size, capable of raising a tree of 15 or 18 inches Diameter. The Costs I am pretty much a stranger to, 15, 20 and 25 Guineas have been spoke of but the price (were it dble that) I should totally disregard provided the engine is capable of performing what is related of it, and not of that complicated nature to be easily disordered and rendered unfit for use, but constructed upon so plain, simple, and durable a Plan that the common Artificers of this Country may be able to set them to rights if any accidents should happen to them. If you should send one be so good as to let me have with it the most ample directions for the effectual using of it, together with a model of its manner of operating."*

Design Reviews

Design review is one of those logical, helpful ideas that should have been adopted long ago but was apparently difficult to accept on psychological grounds. It is seen more and more in industry and works as follows.

At preplanned times during design several more or less formal project meetings are held. A few people thoroughly qualified in one or more aspects of the product are invited. If the project is the design of a special electric motor, for instance, a bearing man, a heat-transfer man, and an experienced motor-manufacturing man might be included. The engineers doing the design describe their work to date and their ideas for continuing it. These contributions are freely discussed by all, and the experts offer suggestions for consideration—perhaps warnings about problems or inadequacies inherent in steps already taken. Or they may offer suggestions for the future or for different approaches.

The success and usefulness of these meetings depend very much, of course, on the ability of all participants to work together as a group to *bring out the best ideas* and to consider them carefully. This is a strain on some project engineers and designers, especially if they think all their ideas must be home grown. But if the project engineer and

* Quoted in *Potomac Squire* by Elswyth Thane, Duell, Sloan, & Pearce, New York, 1963.

his team know that their job is to produce a superior design, regardless of where ideas come from, they welcome the review as a help.

Needless to say, if the reviewers themselves are unaware of the purpose of the meeting and assume an overly critical and holier-than-thou attitude, the designers will have to protect themselves. Benefits of the meeting then are probably few.

Another new idea about design, rapidly gaining adherents, is called *value analysis*. The whole design and each of its parts are examined closely, usually by a team of people. They are examined in terms of *function* and with deliberate efforts at *creativity* and *invention*. The goal is to reduce production costs and increase functional performance. It is basically a very systematic approach to redesign, to improving a functioning design once it has been made and particularly after some experience has been obtained with its production.

Applied to new design work, this technique is sometimes called *value engineering*. It systematically uses design techniques considered in this chapter along with many others. It is especially strong in requiring that all parts of a design be considered from a function standpoint rather than for their own sakes.

Some complain that there is nothing new about these design approaches—that they are but passing fads on the engineering scene. Good designers have, of course, always thought primarily in terms of function and costs and have considered all aspects of a product's life, including manufacture, installation, and maintenance. But these systematic approaches can make most of us perform even better. Perhaps we can profitably take a leaf from the salesmen's notebooks and learn to climb enthusiastically (if selectively and intelligently) on whatever current bandwagon will help us toward our goal—better design.

Good Engineering Practices in Design

1. Recognize the limitless choices included in most design problems. Seek high plateaus of excellence, rather than the last degree of performance.
2. Think out the overall design in terms of function and broad alternatives first.
3. Consider the entire use cycle of the product from manufacture through transportation, storage, installation, operation, checking, maintenance, repair, and scrapping.
4. Identify critical elements in a design early and favor them.

5. Subordinate noncritical elements. Use standard components or designs for these wherever possible.
6. Look for solutions in noncritical areas which can permit either of two risk alternatives in a critical area.
7. Where certain performance requirements prevent an otherwise excellent solution, investigate trade-off possibilities to reduce them.
8. Prove out doubtful critical areas either by analysis or by experiment.
9. Check the resultant design carefully against requirements and interfaces with other parts.

Poor Engineering Practices in Design

1. Seek precise, analytical solutions to general design problems.
2. Fail to take advantage of analytical guides when available.
3. Fragment the design work on a project so completely that component parts do not match.
4. Design so conservatively that performance capabilities are significantly below what the state of the art would allow.
5. Include critical elements which are unproven and will later upset the design.
6. Fail to consider previous designs.
7. Slavishly use previous designs by modification.
8. Fail to make use of standard components.
9. Design with little regard to cost, complexity, producibility.
10. Design without regard to maintainability.
11. Fail to expect changes or to allow for them.

Chapter Nine

Manufacturing and Quality Control

If you are a designer or developer or engaged in sales work, do you really understand the problems of your production group? If you are in production, how much do you know about the engineering work done elsewhere in your company? How does production fit into the rest of the business?

What happens if nobody else understands the producers? Or the producers can't see what the others are doing?

Here is a not uncommon situation that I ran across while in design work. Our design engineering group was responsible for several large equipment projects in various stages of development, design, and production. We were working with a corresponding manufacturing group that built the products we designed. The relations between these two groups, jointly responsible for a large segment of the company's product area, could only be described as hostile. Whenever difficulty arose in production or test, the manufacturing people made every effort to blame it on "engineering"; and the designers tried similarly to show that it was "manufacturing's" fault.

More effort was devoted to fixing blame than to straightening out problems. Earlier the manager of manufacturing had ordered off the factory floor several design engineers who were attempting to work out a solution for some production difficulties. In turn, the design engineers were almost contemptuous of the manufacturing engineers' knowledge and abilities. When I first called on my new counterpart, the manager of the manufacturing group, he was barely civil. Production was almost at a standstill.

Over a period of months this situation was gradually improved through the efforts of the clearer-thinking engineers on both sides. The immediate problem was that *each group was ignorant of what the other's work really entailed* and therefore was unable to cooperate effectively.

114

The work improved as all participants understood more of the total effort, and as the managers involved set an effective example of cooperation. Especially as the attention of both groups of engineers was directed *away from personalities* to the real problems, production increased.

Being an engineer, you will be interested in searching back of the immediate problem. What basic causes produced the original trouble? At this point we will mention two of them.

In the particular company that we have been considering the engineers who manned production and those who manned design were recruited, oriented, trained, administered, and promoted in two completely separate divisions. Neither division paid any particular attention to the needs and interests of the other. Also, production and design functions were separated as "manufacturing" and "engineering" right up through managers who controlled several thousand people. Thus the engineers, whether designers or producers, had very little opportunity or incentive to develop understandings and attitudes which could induce a really effective cooperation.

In small companies you will have the advantage of dealing closely with both design and production. But as a beginning business enterprise grows, it first separates out the production function. Presumably this makes all groups more efficient in that they can concentrate as experts in particular areas of work, and such a separation is probably useful for most manufacturing and some other kinds of business. The experience above, however, hints at severe disadvantages that you will have to recognize and overcome.

We will devote this chapter, then, to understanding the relation between production and design in a modern enterprise. You remember that these are steps 4 and 3, respectively, in the engineering cycle. Although we use a factory as a specific example of production, you can translate these ideas into construction or process work as you like.

Production Considered as Flow

The technology of production is extensive and well covered in many textbooks. It expands rapidly, and its advances are chronicled in numerous production-oriented journals. In many respects the technology of production is more systematic and better understood than that of conception and design. Both depend, of course, on the same basic physical technology.

Design will be catching up, from a system standpoint, in the years to come. At the same time it appears that great revolutions (automation

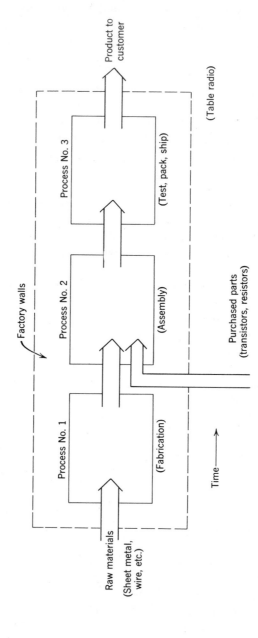

Fig. 9-1. Elementary physical factory flow production is basically a flow of materials and assemblies. Consider a table radio set as an example here.

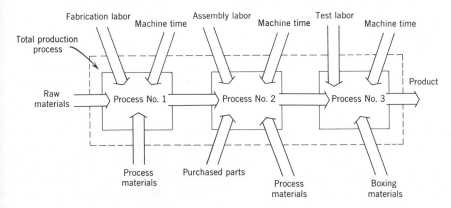

Fig. 9-2. Production flow concept which considers more of the essential inputs.

is just one) are immediately ahead for production. Production technology has enjoyed a surprisingly unhurried and steady rate of advance until recently.*

We are interested here, not in the details of production technology, but in its broader philosophy. The concept of production is that of *flow* or, better, *controlled flow*. Figure 9-1 shows physical raw materials and purchased parts flowing smoothly into a factory. They move easily from station to station, where they are processed by various machines or people. Via testing, packing, and accounting stations, they flow as products out of the factory onto transportation means and so to the ultimate consumer.

The construction (production) of works such as buildings, dams, or bridges has a similar flow except for the fact that the product remains on the spot and waste materials are removed (as they are in factory production). This general pattern has some variations for different industries and factory set-ups, but they will not affect our reasoning here.

If you want to take a somewhat more sophisticated look at production, your flow will include (with the physical materials and product) such other contributions as labor and machine operations and test and inspection operations. That is, these elements of the more or less continuous factory process will be thought of as entering the flow and combining with physical elements to make up the outgoing product (Fig. 9-2).

---

* I suspect that conception-design and production are going to become increasingly hard to distinguish clearly in some areas. Perhaps we may differentiate on the criterion of certainty. Conception-design is by its nature a highly probablistic activity. Production should be much less so.

Information Flow

For a yet more advanced understanding of manufacturing flow you must recognize the existence of a parallel and closely related information flow. This touches the production flow at many points, contributing to it and taking from it.

This information flow controls production. It starts with design information, drawings, materials specifications, process specifications; it continues with detailed manufacturing orders and instructions worked out by the manufacturing planning personnel; it includes information on process changes. Shortage information, test results, and troubles come from the process back to the control flow.

Figure 9-3 attempts to depict some of this interrelation between the two flows. The time relation is quite approximate. There is much recycling. With this concept of production before us it is easy for the conception-design engineer to understand at least the broad outlines of producer activity.

Engineers in production are interested primarily in establishing a *smooth and efficient flow* in both the production channel and the information channel. To this end some of their typical and often specialized functions are the following:

Planning:	Designing the information flow system to control production.
	Deciding in detail how a part or product will be routed through the manufacturing facility, what stations must be established, what machines used, what kind of workers and how many will be needed, etc.
Manufacturing engineering:	Designing the production machinery and set-ups.
Procurement:	Purchasing or otherwise providing for the flow of materials and parts into the production flow from outside. These must be of the right kind (meet specifications) and in the right quantity, and must arrive at the right time.
Quality control:	Planning the test and inspection system, designing test station set-ups and machines, operating the testing system.

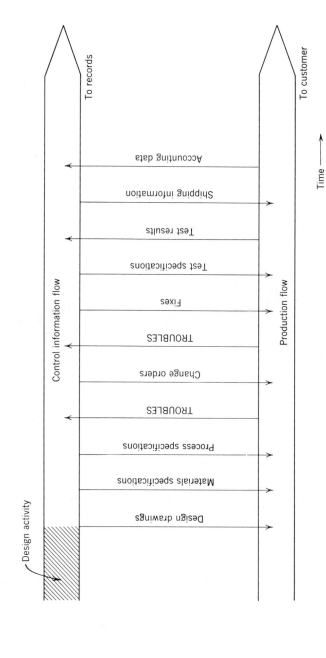

Design activity

Control information flow

To records

Accounting data
Shipping information
Test results
Test specifications
Fixes
TROUBLES
Change orders
TROUBLES
Process specifications
Materials specifications
Design drawings

Production flow

To customer

Time ⟶

Fig. 9-3. Relation between production control information flow and production flow. There is a controlling flow of information that parallels the actual production flow and interacts with it at many points.

Supervision: Exercising general control and integration of the above functions, particularly of the fabrication and assembly operations themselves.

Production and Business Success

Profit and loss are the basic measure of how successfully engineering work is being performed, that is, of whether human needs are being better met by a company's activities. Presumably, if the devices for sale are indeed a contribution to meeting some need, the company will prosper.

There are of course considerations like advertising, other sales effort, location, and competition that might appear to modify this conclusion. But careful thought will show that most of these factors bear directly on how useful the company is to its customers. Any other minor influences should balance out over a period of time. Such factors serve also to emphasize again the need for the engineer to think of his work in its entirety.

The usefulness of a product in fulfilling human need is generally some inverse function of its cost. Thus, for a given product, *the engineer organizes production to minimize cost.* This task falls largely at present on production engineers, though all must participate. It is the producers who are held generally accountable for financial success in many enterprises after a design has once been made and put into production.

There is usually a certain amount of injustice in this practice. Design is the major determinant of product cost in the first place, although some engineering managements seem to ignore this fact. The designer can thus understand some of the sensitivity of his colleague in production.

From the ordinary producer's standpoint, an ideal business situation is one in which a standard product is made with little or no change over many years. The producer is thus free to optimize his production flow, improve his methods at all points, and calculate costs of new machines. He can do all this quite deliberately and thus with considerable efficiency and confidence. As production costs are accurately broken down, value analysis methods may be used to feed back information for improving design. These improvements can be integrated gradually into production for even better costs.

Traditional production attitudes seem to be based pretty much on this kind of ideal picture. When the designer is unable to provide all of his specifications in a timely way, when production runs become short and changes frequent, or when the designer's specifications are inconsistent and result in unpredictable rejects, the producer's orderly world is upset. He resents this disruption, especially when the front office is looking to him for a profitable venture. He becomes critical, suspicious, and uncooperative.

Furthermore, even in the best of circumstances, a producer has many other problems that seldom plague the designer. He usually deals with large numbers of workers who have their own problems and individualities. Often many of these people are organized into legal bargaining units, with consequently divided loyalty and touchiness. The producer deals with tight schedules many factors of which he cannot easily control—his vendors, for instance. Because of these problems and because of the immediate importance of production to the business financially, the production function is usually separated early from the rest of a company.

The designer's picture of business flow is somewhat different from the production flow discussed above. He thinks more in terms of a sequence of the five engineering functions: conceive, experiment, design, build, test. He tends (unreasonably, as it turns out) to regard the building or production function as more or less of a detail. He shows the producers how to do it anyway, he supposes. He continually looks for new ideas, radical improvements in the design.

Unfortunately designers are often not as cost conscious as producers. Design costs are a one-time item for any product at worst, and often design engineers are carried on overhead. A designer sometimes misses the point that genuine functional design improvements can be so expensive in production that they result in no help to the user at all.

But the picture is not all one-sided. If the producer is allowed to continue indefinitely with the same product and little or no improvement, he will become progressively less useful as newer ideas and competitors pass him by. In plain words, the company will go bankrupt. (In those situations where little improvement is possible, he had better turn his duties over to a more clerically minded individual and go on to professional work elsewhere.)

As in the example with which we began the chapter, the cure for these attitudes lies in *learning more about the entire engineering process.* Neither designer nor producer can effectively do his part of the work unless he performs it consciously as part of the whole process of applying technology to better meet human need.

That designer has not yet engineered whose designs are not being produced and used. The producer is not engineering if he is merely going on in the same way with no improvements or readiness for the next step. Excellent engineers in both functions, recognizing the ultimate purpose of their efforts, work smoothly and intelligently together to optimize the entire need-meeting operation.

Production Functions

There is more routine work in production than in conception-design, and so there are many places where an engineer carefully provides for delegation of this routine to clerical personnel and computers. He thereby frees himself to pursue the more professional task of seeking improvements. In cooperation with the designers he looks for improvements in the product for production purposes. And he looks especially for improvements in production processes themselves, including organization.

Manufacturing engineering—the design of production equipment and set-ups—is no different from the design that we discussed in Chapter 8. In some ways it is easier than the design of a product to be sold, since its immediate object is clear and apparent and the customer at hand. Manufacturing engineers pay particular attention to life requirements. Often this equipment is needed on a temporary basis and should not be overdesigned. Particular attention is also given to the possibility of designing production and testing equipment in such a way that it can be inexpensively modified for other jobs. Appearance design is more flexible than with a product, although it is not unimportant. Unusual combinations of purchased and made parts can often be used advantageously.

The *production planning* function (including information flow) designs the production scheme for a product to include such items as materials specifications, procurement, inspection, storing; machines to be used and processes on each; material-handling procedures at each station; control methods to be used, including forms and procedures or deviations from standard organization procedures; personnel requirements, including training. This kind of work, which can be very creative and professional, requires a strong background in production technology.

Procurement or purchasing is a major function of production since it will have decisive effects on production costs and production flow. It is often done by nontechnical people who have some trouble communicating with engineers. After a product is once in successful production, most of the floor problems that designers are called on to help solve arise from vendor situations.

Here is a typical problem. The purchasing people find a vendor who offers a lower price for some item inadequately specified by designers or production planners. The first shipments from the new vendor enter the production flow and promptly stop the line. Designers find the trouble and complain to procurement about what has been bought. The buyers refuse to change the vendor since he is meeting established specifications and they are proud of having saved money by finding him. Sooner or later, some intelligent engineers have to get together and resolve these differences.

Quality Control

Quality control (QC) is ensuring that the product as it comes from the production process meets the specifications required of it. Quality control engineers set up testing stations in the production flow to pass or reject the item tested. On some types of product tests are made by sampling.

Testing only the final product usually results in an intolerable rejection rate because of relatively simple difficulties early in the production flow. Therefore other QC test stations are established to examine and pass parts before assembly. You will find interesting problems of design in the entire test program for a product.

Quality control is certainly a part of the manufacturing function since its purpose is to make sure that production work results are satisfactory. Test stations in a production flow are basically no different from those stations that obviously modify the product in some way. Test stations also modify the product in the sense that items coming out differ from those going in because now more information is known about them.

There have been cases, however, in which short-sighted production people (interested only in today's goal of so many units through the factory) were unwilling to accept the responsibility of ensuring that what they produced would be within specifications. Hence in many organizations the QC function is separated from actual production to provide an independent check on the quality of manufactured product.

I once had a customer who nearly balked on a large production contract because our QC organization reported to the manager of manufacturing. He believed that, if we did not reorganize so this function reported directly to the general manager, test data would be so warped under production pressure that he might receive poor equipments.

But such functional separation leads on occasion to ludicrous attitudes between QC engineers and other production supervisors. Fabrication or assembly leaders, harried by management for more output, come to regard the QC test stations are nefarious traps which they must by-pass. Less frequently the QC people may also feel that their duty is to make others toe the line by continually tightening quality requirements.

Again, the solution here is for both to look at their overall, need-meeting goal. The producers then will not want to let anything go out of the factory which cannot accomplish the established purpose. Testers will want their quality requirements to be as loose as possible while still consistent with product specifications.

Unfortunately things are not quite that simple. Many modern products are so complex that there is no direct analytical way to derive intermediate quality requirements from product performance specifications. All the engineers involved in the design, production, and testing of a product must cooperate to find realistic intermediate testing specifications.

Relation between Quality Control and Reliability Engineering

Quality control engineering (whose function, as we have seen, is to ensure the *quality* of the *product produced*) is closely related to reliability engineering. The function of the latter is to predict and improve the *reliability* of the *product designed,* that is, to determine how long it can be expected to carry out its intended purpose. Quality control is most logically a function of production, and reliability a function of design.

Poor QC adversely affects the reliability of the product, but good QC cannot improve it. Either reliability is designed into the product in the first place, or it will not be there. In one sense every design engineer should be a reliability engineer and design proper reliability into everything that he turns out. Of course, there is usually an optimum reliability when traded off with cost.

In the same sense every production engineer should be quality-minded and work to ensure that his part of the production process is contributing to a quality product and not to rejects and problems. Poor design can certainly adversely affect the production quality of a product, but good design cannot make up for poor QC during production. Unless quality is built into the product by the producers, it is not there.

Trouble Shooting

Ideally, when designers complete drawings and specifications on a new product, they have taken into consideration the needs of producers and have configured their designs to be easily and inexpensively made. In practice this ideal state is never attained and cannot even be approached without a great deal of recycling between the building and designing steps of the engineering cycle. Thus designers must *follow their work right into production* to find quickly what has to be corrected at once, what else should be modified, and what can be improved in the next design.

Whatever stops or seriously impedes production flow or seriously jeopardizes product quality must be corrected at once. This trouble shooting on the (factory) floor by design engineers is a fascinating part of the business. It requires skill and is obviously done under great pressure. Once a production line has been started, it is expensive to stop or delay manufacture. The machines have been designed and put into place, operators hired and trained, and materials and parts ordered and received. Often it is desirable to make pilot production runs first to minimize these difficulties, but they are expensive, too.

In addition to possessing normal technological skills a good trouble-shooting engineer becomes adept at quickly isolating trouble. It may be found to be a drawing error or some more fundamental shortcoming in design. It may be an inadequacy in a production machine or production test set-up. There may be some vendor difficulty, or a training problem among the operators. Whatever it is it cannot be corrected until found.

Jealous attitudes on the part of designers, fabricators, or testers at this point only cause delay in finding the trouble. When a new line is first set up, there may be multiple problems intertwined with each other. After a line has operated satisfactorily for a while it is almost always possible to trace any sudden problem to a single cause. Such a cause may have multiple effects.

Once a trouble has been tracked down and isolated, its correction is also a matter for intelligent, alert cooperation. Although the difficulty may be a design error, the most inexpensive solution may be a change in some production procedure. If the trouble is a vendor problem, it may in some instances be cheaper to modify the design to accept the part or material variation. Some revisions are used on a temporary basis until a more fundamental correction can be made.

Sometimes designers make the mistake of thinking that, if there are no loud screams from producers—that is, no problems which have to be corrected at once—there is nothing further for them to do after a design has been released to the factory. The experienced designer, however, watches initial production closely for (a) indications of changes to be made before some serious or costly breakdown occurs,

(b) other changes which could reduce production costs or improve the product, and (c) design improvement ideas which it may not be advantageous to make at this time, but which should be collected for the next design or modification of the present design.

In general, if there is substantial doubt about whether a change should be made it is not made. Smooth production flow is too important to a business to be interrupted without strong reasons.

Good product design (and to a slightly lesser extent good manufacturing engineering and QC design) should not require excessive changes after production starts. The number of modifications required is a rather good inverse measure of design excellence. In "state-of-the-art" industries like military electronics there may be some excuse for a few more modifications, but even here design should require a minimum of changes from recycling.* This fact is, however, no excuse for either designers or producers to avoid changes or to ignore the requirement for them. Difficulties must be quickly diagnosed and cooperatively resolved.

Once production is smoothed and well under way it is a good idea for both producers and designers to appraise (on paper) the design from a production standpoint so that lessons useful to the next design will not be lost. Designers should take the lead in this evaluation since producers will then be busy with everyday problems.

Good Engineering Practices in Manufacture and Quality Control

1. Designers and producers jointly and periodically review the progress of the project from its inception through the attainment of smooth production.
2. Producers take the earliest opportunity to begin definition of production needs in processes, equipments, and machinery design.
3. Designers make adjustment of designs to facilitate manufacture and quality test. They seek information and assistance from production people.
4. Designers establish test criteria as loosely as possible while still meeting component and product requirements. In some cases the inclusion of one or more "adjustable-in-production" components will beneficially relax procurement or production problems elsewhere.

* I remember a production meeting where a document pertaining to the job was being reviewed. A typist had made an interesting error, rendering the words "designed and in production" as "designed in production." This caused quite a bit of ironic merriment since the job had required a great many changes after reaching the production floor.

of one or more "adjustable-in-production" components will beneficially relax procurement or production problems elsewhere.

5. Designers and producers keep each other informed as early as possible concerning the need for or prospect of design changes. While these are held to a minimum, they are treated by all as a normal part of the production cycle to be handled as expeditiously and efficiently as possible.

6. Quality control test specifications are established as basic and independent wherever possible.

Poor Engineering Practices in Relation to Manufacture and Quality Control

1. Little or no regard is given to manufacturing and test problems until the design is completed.

2. Because producers do not work out at least broad processes early, designers are unable to minimize production problems through improved designs for manufacturability.

3. Designers and producers look at a project as more or less completely divided into design and production phases with no time overlap and little connection.

4. Designers fail to release partial information as early as possible to permit early producer planning.

5. Designers permit producers to badger them into premature release of very tentative information, wasting producers' early efforts and tending to box in the design.

6. Component QC test criteria are set which do not reasonably ensure that assemblies will meet their specifications.

7. Worst case design and test criteria are carried to a point where component tolerances are ridiculously costly.

8. "Functional" QC tests are used to such an extent that no workable understanding or control of the product is available.

Chapter Ten

Research and Development

All engineers are changers and improvers, by virtue of their professional mission. If you formally devote most of your time and effort to looking for useful ways to change, however, you're a development engineer or possibly an engineering researcher. These people seem to become more numerous every year. In one large technically oriented company about three-quarters of their several thousand engineers are in "R and D" work.

Two chapters back we saw that most engineering activity is built around the design function. A designer continually needs new, creative ways of accomplishing his purposes. A development engineer's function is to provide new ideas that the designer can use.

But it is not enough for a designer to have new ideas. He must also have sufficient confidence in them to be reasonably sure that his design will work when they are included. The step from a new idea to enough confidence to include it in a design is often a long and expensive one.

Turbine engine automobiles are a good current example of engineering development work. Are they really feasible? Many problems must be solved in order to design such an engine within the constraints of today's automobile market. Beyond the engine itself, what are the performance characteristics of a car so powered? How will the using public react to it?

Large automobile manufacturers have spent millions on the technical problems in the last few years. Recently one of them began to put a number of trial cars into the hands of consumers for extensive tests.

That kind of activity, aimed at *first working out the feasibility of an idea to see if it is possible within significant constraints, and second finding some of the better ways to use it if it is feasible, is development.*

Looking at our five-step engineering cycle broadly, we see that development work is the experimental stage; but it is not, of course, limited to experiment in the laboratory sense. Development projects are set up to establish the feasibility of new product ideas. "Product idea" here could include process or procedure or any other contribution to meeting human need.

128

Similarly *engineering research* is the conceiving step, creatively recognizing a new idea. It is *looking into nature, usually physical science, to see what possibilities exist for better meeting a human need.* It may be looking for a way to better design some device or process or structure. Or it might even be seeking a better way to design some analysis method or computer programming scheme.

So you can see that the sequence goes as follows: A researcher looks into a certain area of scientific advances to see what possiblities it may have for product improvement in his company. He comes up with several ideas. Developers take the best of these ideas and work further on them to show that they are really feasible for design and how they might be used. Some of the ideas make the grade here, and some don't.

You know the next step: designers use the proven developmental ideas in their new designs. If the turbine engine car turns out to be reasonably satisfactory in the hands of users, no doubt such engines will be designed into the regular line of cars. Designers profit by both the strong and the weak points of the development project.

Observe once again an interesting property of the five engineering functions—that they can all be applied to any one of them. In the conceive, experiment, design, build, test cycle, development work—experiment—like any other engineering work, will involve all functions.

Again taking the developmental turbine engine car as an example, you can see that as a developmental tool it certainly had to be conceived; it was probably experimented with both on paper and also by building earlier models; then the final developmental version for consumer trial had to be designed and built. Also, it must have been tested to some extent before being placed into the hands of trial consumers. This present stage of the developmental process was surely arrived at through the recycling of earlier developmental steps.

Now there is no clear line between engineering research and development, or between development and design. These two boundaries are not really defined at all. We know what the differences are between their centers, but we won't waste time trying to define boundaries clearly.

The designer invents a little, takes a *few* risks on his own, but he has to be quite certain that his design will work (or at least that it can be made to work). Good developers have one or more reasonably sound design approaches started before they turn work over to design colleagues.

Similarly developers can't help doing some research as they go along. They have deliberately opened their thinking to new ideas and creative inspiration. And the researcher must develop his new ideas, at least slightly, to tell whether he has found anything yet. A pretty good rule of thumb for each of these engineers is to push ahead rather than to hold back.

Scientific Research and Engineering Research

In thinking carefully about engineering research and development it is essential that you differentiate it from scientific research. Many inadequacies and shortcomings of engineering R and D work arise from confusing science and engineering. Let's take another quick look at the fundamentals.

We concluded on looking at this difference in Chapter 2 that science was an effort to establish a systematic body of knowledge about nature, using the idea of "nature" in a very inclusive sense. Nature is the total environment, including ourselves. The scientist is interested in finding out more about nature and relating it to what he already knows.

We engineers, on the contrary, are interested in applying technology—which includes scientific knowledge—to do something useful better. Engineering is the application of technology to meet human need more effectively.

Now, engineers have a strong interest in science. Because of the technological base of our profession we have to. Science gives us new basic tools and many of the ideas that we need to discover strategic factors. But that doesn't make science engineering.

People who don't distinguish between these conflicting goals can do some pretty poor engineering development work on occasion. The difference is primarily one of attitude. The scientist is trying to learn; the engineer is trying to do. It's almost as simple as that. They do many identical things. But when the scientist is doing, it is in order to learn. When the engineer is learning, it is in order to do. Don't mix these attitudes up.

If you want to be a scientist, go be one. We need lots of good ones. We engineers owe a great deal to the scientists and hope to owe much more. But don't confuse science and engineering, as many young engineers seem to do today to the detriment of their work. As an engineer you will drive through to some useful result in your work, but only if you keep that goal constantly in mind. If you mistake your work for science, you may be satisfied with just learning something new—or trying to.

Now carry this difference into research work. Scientific research is aimed at learning new things and relating them to the general scheme. Engineering research is aimed at learning new things for the purpose of using them. Undoubtedly the scientist hopes that his work will be useful, too, but his purpose is to fill in gaps of knowledge. He knows that his kind of research is most useful in the long run if he makes that his goal.

The engineer is not satisfied in development work or engineering research with merely learning things. He constantly thinks, "How can I use this?" "What is limiting my application of that fact?" The scientist is well satisfied with negative results. Although these are useful to the engineer, he would consider his overall development effort (and the research preceding it) a failure if it simply proved that an idea can't be made to work. If this is going to happen to you, make it happen fast and then go on to something else.

Probabilistic Nature of Development

When an engineering idea reaches the *design stage* you have great confidence that it will work, that it is quite feasible. Rarely should an idea get as far as design and then be found impractical. Although all the *details* are not fixed—the designers will work these out to advantage—it is expected that, if one idea for a solution detail will not be adequate, another can be found to take its place. Thus the whole project idea proceeds to its ultimate successful conclusion.

On the other hand, R and D is by its very nature more probabilistic. You have no real assurance that a development idea is going to work out at all. Certainly the way in which it can be made to work is not clear. *It is this strong element of chance in development work that best distinguishes it from design.*

We have seen above that a major source of trouble in development work is to proceed, perhaps inadvertently, as if it were a scientific investigation rather than a results-oriented, engineering effort. A second and equally serious source of difficulty is failing to recognize and provide for the chance nature of the development project. This point is so important to you in R and D that, at the risk of boring a few of our more sophisticated readers, we will stress it in some very simple ways.

"Probabilistic" is an adjective that describes the noun it modifies as being subject to probability rather than certainty. For instance, in throwing dice the resultant number is probabilistic in that it is subject to chance and cannot be foretold with certainty. (Of course this is presumably the intention in that kind of game. If the results were predictable, some of the players should take action at once!)

Another example is the fall of an artillery shell. A cannon is carefully adjusted so that its shells should land on a certain point. But the shells will instead land over a small area whose center, if the adjustment is correct, is the designated point. Some will fall beyond the point and some short of it. Hence, although we are quite confident that any one shell will fall within perhaps 200 yards of the target, it is a matter of chance exactly where it will fall within the 400-yard zone about the center. The only way to know where it will fall is to fire it and see.

Or suppose that a cylindrical part is being manufactured on an automatic machine to a nominal diameter of 3 inches. All engineers will recognize that the exact diameter of any piece before it is measured is probabilistic. Even with the automatic lathe properly adjusted there will be a slight variation in the diameters produced. Of course any given piece after manufacture has a specific diameter that can be found by measuring it.

In an exactly analogous way the outcome of R and D projects and of most parts of projects is probabilistic—subject to chance. You can't tell to begin with about success or the form that results will take or the avenues down which development will lead. If you could predict these things, there would be no development work. The idea could go right into design.

As an example, consider the time when the tunnel diode was a relatively new device. Suppose that the chief engineer of a radio receiver company had given one of his project engineers the job of looking into the possibilities of tunnel diodes. Let us assume that the chief decided to allot $25,000 and 6 months for this project. Presumably it would tell how, if at all, the tunnel diode could be used by the company at the present time and in the immediate future. Could it be used to improve the performance of receivers? Could it be applied to reduce costs or improve maintenance or reliability?

Surely you would feel at the beginning of such a project that no one could certainly predict how it would come out. It might develop that, interesting as these devices are, there was no possibility of using them in the next few years. On the other hand it might be that you could reduce costs by 25% and forge ahead of the competition. Perhaps there would be more modest possibilities and consequent difficulties in deciding whether to design with the new device now or wait for improvements to follow.

Just as the diameter of the automatically machined cylinder could not be predicted certainly over a significant range of values, so the result of this development project is similarly uncertain. It might turn out that the device would be useless for amplifier circuits but could be advantageously employed for oscillators. It might be good for low-level stages of receivers but not for the output stages.

Evidently the project engineer would be foolish to go into his work with a fixed idea of what was going to come out. Some other result might turn out to be more practical. At any rate the results of this kind of development project are widely (or even wildly) probabilistic.

Contrast this again with a design project for which the outcome is reasonably certain. Although there is still some risk in details here, as in almost any engineering work, probable errors, in a manner of speaking, are kept much lower than in R and D.

In fact, in an original design there is usually so much at stake that it is common to "overdesign" to be certain the product will perform acceptably. After successful production experience redesigns can be made on a much less conservative basis. The designer has control of the variables involved now and can retreat toward the old design if necessary.

For the very reason that design must be well assured of success, the engineering manager first works out chancy possibilities for design improvement with a development project.

Engineers who do R and D, particularly those who lead these projects, but do not fully understand the probabilistic nature of the work get into two kinds of trouble. First, they lose control of their projects almost at the outset. Second, they fail to uncover the real technical possibilities that exist.

Conduct and Control of Development Projects

Careful Control Indispensable

Precisely because R and D projects are probabilistic to a high degree and therefore difficult to control, they need the most careful and exact kind of control. All general control principles and techniques discussed in Chapter 4 apply here. You use them to maximize the probability of success and to minimize losses.

We have already observed the attitude (in Chapter 4) wherein an engineer, overwhelmed by the chances he must take in development work, denies the possibility of any effective project control at all. This is like saying that if your team of thoroughbred horses is spirited and hard to handle you should just let it run off with you. This helpless attitude makes about as much sense in an R and D project as in a horseshow.

Let us look again at some of our control techniques, with special reference to R and D.

Use of Critical Factors

It is especially important to *identify the critical factors early and often.* Often, because in a free-wheeling development project these factors change rapidly as you advance and learn new things. Early, because in the nebulous areas of R and D it is easy for you to waste time and money going in the wrong direction or working on the wrong problem.

Having identified the critical factors, *put essentially all your effort on them,* subordinating everything else. Although this concentration is important to a degree in all work, it is doubly significant here, since the whole idea of R and D is to break through to new possibilities. The following design work can often take care of routine details, anyway. (Be sure, of course, that what you leave undone is really a detail and not a hidden trap!)

If a critical factor is not strategic, cannot be broken through, it is wasteful to develop the rest of that solution. For example, if the applicability of tunnel diodes to oscillator circuits had appeared to depend primarily on temperature characteristics, that factor should have been investigated first. You would not waste effort on the less critical circuitry until the temperature characteristic was found to be usable.

Pilot Programs

This same idea is often expanded in another way. Perhaps a big effort is needed in a certain area, say the general area of tunnel diodes as amplifiers, to go back to our example again. It is common to *first mount a relatively small and financially modest program* to probe the more critical limiting factors. If this is successful, you can then proceed with the full-scale effort. *Pilot programs* give you further insurance against pouring large amounts of effort in worthless areas of work ("ratholes").

Stop When You Are Done

This last technique emphasizes again a familiar point: *stop a project or subordinate effort in a timely way.* Watch the R and D project carefully. Don't discourage easily, but stop the project or parts of it when they have gone far enough to show that probably little can be gained by continuing.

Project engineers on this kind of work are particularly alert to *redirect or reorient the project or parts of it.* When one promising approach turns out to be rather clearly unfeasible, they turn the group effort into more promising channels with minimum loss of funds and momentum. Reorientation must consider the funds remaining. It is a rare development project that will ever run a course close to what was first planned. Redirection and change are the essence of R and D.

Keep Your Thinking Flexible

An engineer who does not appreciate this probabilistic nature of R and D is unready to change and twist with the project. He is less likely to achieve the best attainable results in his work. Each member of the team has to *remain* as *flexible and open-minded* as possible concerning how the project will go. If you hold a minimum of preconceptions on how things are supposed to proceed, you are willing to turn in new directions for the answers you need.

Ideally a project will keep *several parallel solution ideas* for each critical element. To habitually have available a second idea before the first fails avoids a morale problem. A very helpful attitude here is to keep the *ultimate project goal* in mind and to refuse to be bogged down by troubles with intermediate aims.

State of the Art

In the design discussion of Chapter 8 we covered the proper and improper use of the old design as a starting point. The corresponding situation in R and D is the use of what is known as *"state of the art."* This term is employed in slightly different ways, but we will take it to mean what is *currently* known or what is the *latest current information* on the subject at hand. The state of the art in tunnel diodes, for example, is changing rapidly and will not be found in any textbook.

Engineers interested in a particular area try to keep up with it through journals and meetings. Naturally R and D is done in rapidly advancing areas where it will almost always be difficult to know the state of the art. If you don't know the state of the art, you may be reinventing the wheel, that is, repeating something that has been done already, thus wasting time and money. Conversely, if you spend two-thirds of your project allotment in learning the state of the art (by library research or other means), you will not get much new work done. Researchers and developers try to strike some kind of balance here. If you keep up pretty regularly with your area of specialization, the problem will be easier.

Don't Be Conservative in Research and Development

The old proverb "nothing ventured, nothing gained" should be displayed in neon lights above the desk of each development engineer.

There is much about engineering that has a slightly conservative flavor, in the best sense of the word "conservative." When an engineer gives an opinion, he speaks from a real physical understanding of the situation or not at all. The designer is sufficiently conservative so that his design can be expected to work with little trouble. You could suggest other examples of proper engineering conservatism.

But *conservatism in R and D goals is inadmissble*. Oddly enough it seems to go hand in hand with uncontrolled projects. Both sins rest mainly on misapprehending the probabilistic nature of R and D.

Sometimes an experienced design engineer is unable to shake off his conservatism when he comes into development work. Not infrequently an engineer who has been associated with a poor development project becomes quite conservative on his next one. He has been burned, he thinks, by the probabilistic nature of the work. More probably he was burned by his own poor control procedures.

Success in R and D work involves breaking through in some essentially unknown area. By conservatively limiting his efforts to what can be foreseen (or at least dimly perceived) the engineer literally shuts his eyes to the real possibilities. Here, as elsewhere, all risk taking is done as judiciously as possible to maximize gain and minimize losses. This means especially careful attention to critical factors. But *risks must be taken*.

The best simple comparison I can think of for risk taking on a well-run R and D project is seen in the Sunday afternoon work of some professional football quarterbacks. The costs of risks are carefully considered, but they are taken freely when appropriate. And as the period draws to a close the quarterback will boldly call a long touchdown pass instead of grimly inching down the field on the ground. His play often succeeds. Even more frequently his team will stand high in the season averages.

Good Engineering Practices in Research and Development

1. Build a reasonably thorough background of information on what has already been done before attempting to seek new results.

2. Deliberately assume technical risks on a job in order to obtain technical returns from the work.
3. Keep two or three parallel approaches in hand, particularly for each critical area.
4. Devote as large a part of your work to critical project areas and as small a part to noncritical areas as possible. Use off-the-shelf components and techniques for noncritical areas wherever feasible.
5. Attack critical areas and the most promising of parallel approaches first, before expending large efforts on other parts of a project.
6. Evaluate progress continually in terms of goals sought and means remaining, rather than in terms of point of start and means expended.
7. If results are turning out almost surely negative, take timely action to redirect or discontinue the project.
8. When success begins to be well established, prepare to exploit it in several directions.

Poor Engineering Practices in Research and Development

1. Confuse engineering applied research with basic scientific research.
2. Propose and undertake R and D work without a careful analysis and frank acknowledgement of risks involved.
3. Undertake a complete R and D program aimed at eventual product design before broad feasibility has been established by study of the critical elements.
4. Take so conservative an approach to what can be expected that promising avenues of solution are unexplored.
5. Allow the entire success of the project to hinge on the feasibility of one technique or device, when alternative approaches could be considered.
6. Lose sight of the purpose for which your work is being conducted; thereby develop useless solutions or overlook widely different solutions which could accomplish the end sought.
7. Fail to consider all possibilities because of conscious or unconscious bias toward certain solutions.
8. Allow the financial sponsor (either inside or outside the organization conducting the work) to lose sight of the specific risks involved.

Chapter Eleven

Studies

You will frequently be called on to help with studies. Understanding the advantages and limitations covered in this chapter will enable you to make the most of the time and money available for this kind of work.

Engineering work which involves construction or experimentation with hardware or structures is almost always more expensive in time and money than paper investigations of the same problem. Once an engineering hardware project is begun, it is less flexible than paper analysis and study. More is committed at the outset. To stop short or change will usually make results obtained to that point very expensive.

For this reason it is customary on large development projects to begin with a study phase even though hardware work is essential later. The study *is aimed at raising the level of confidence in the eventual success of a full-scale development.* Because of its simple flexibility the study can cover a wide area quickly. It can eliminate poor ideas and identify one or several approaches which appear to have real potential. Alternatively, results can be negative, indicating that no further development investment should be made.

Studies are also used as the entire development phase in situations that preclude any significant hardware experimentation. Examples are buildings, dams, and other complex systems.

Proposals for doing development or other engineering work are principally prepared as studies. Studies can be used to fill in ancillary areas of large hardware projects where experimentation is thought unnecessary. The more expensive an overall project will be, the more extensive the studies made. The possibility then exists of saving a great deal through even small improvements in approach.

The usefulness of studies, however, is limited. They are only paper creations. They do not provide the realism and the sobering checks that engineers get from building and operating hardware.

We saw the engineer's concern that all significant factors be really taken into account. After design and construction the test function tells him that nothing has been omitted. Before design and construction the experimental function—particularly laboratory or hardware experiments—provides a partial preview of the later tests and increases to an adequate level the engineer's confidence in his design. Development studies, however, are deficient in this respect since they include little or no practical experiment. Their conclusions are always suspect.

Two common shortcomings encountered in study work are (a) neglect of physical fact, and (b) refusal to make such progress and attain such results as can be had (even without physical checks). In making the first mistake some engineers move into a dream world of impossible paper components and unrealistic mathematical models. They are not guided, unfortunately, by the wholesome presence of hardware reality. The second mistake occurs when some engineers, frightened by their inability to obtain physical checks, become extremely conservative. They limit their work and its possible results.

In the excellent study, then, you take advantage of the typical freedom of movement available while recognizing the danger of limited hardware checks. You present your conclusions carefully and lucidly, showing how much confidence is to be attached to each. The way in which these confidence estimates are made should be clear, also. The reader can then judge for himself the validity of your conclusions.

How to Keep on the Track

In a design or development project hardware focuses the effort of all participants and integrates the work. In a study project you must substitute artificial centers of interest. Announce clear goals for the whole study.

There is a tendency for people with little experience in studies to "study an area." For example, on the project for the radio manufacturer discussed in Chapter 10, a preliminary study before undertaking experimental development might be erroneously aimed at "studying the tunnel diode." But this is not what is really needed at all.

A better goal would be "decide whether the tunnel diode can be profitably applied to our product line and if so how." Such a goal immediately starts the engineers thinking in a practical direction. It suggests an initial breakdown for the work. It asks a question that each participant realizes his results will have to help answer. It almost outlines the final report by itself.

In study work don't be caught just "studying" something! Study to answer one or more specific questions, and be sure that these questions are really significant.

This is also an excellent way for you to carry on your day-to-day creative work. Break tasks down into a series of questions. Answers to some suggest others. Always keep the overall goal in mind. Where job notebooks are kept, the question approach can be carried on right in the notebook. It is helpful too in project memoranda. In project meetings a series of specific questions on a chart or blackboard can serve to focus the attention of the whole group.

To take the place of a piece of hardware, the project engineer and his associates think in terms of a specific answer or answers coming together gradually as a result of the work. To enhance realism here the project engineer can occasionally circulate a "today's solution" (or maybe a dummy of the final report) which attempts to give an answer for the whole project. He puts in blanks where information is lacking. Or he may give very tentative conclusions and an indication of why they cannot yet be accepted. As the project progresses, these summaries gradually become more complete and factual.

Such a summary document enables each team member to fit his effort into the whole project. It guides and encourages him. He will want to be sure, though, that it does not freeze his outlook and stultify his creativity.*

It is often not impossible to use a limited amount of laboratory or other experimental work even on a study. By simulation, analog and digital computers can provide quite realistic results for some situations. Often they offer the only practical means to "experiment" with large systems and structures. However, they have their limitations. You'll want to be especially careful not to lose sight of the limitations of your "model" after you get it on the machine.

Control of Study Projects

In some ways it is easier to control a study project than a hardware project. Mistakes are not so costly, and it is easier to redirect the work. It is usually possible to probe unlikely avenues of approach a little farther to be sure that they are not profitable.

In other ways, however, the control is more difficult. We have observed that the integrating effect of hardware is lacking. Most engineers can easily visualize a physical device and understand their part in creating

* Project engineers have been known to insert in these summaries exactly the opposite conclusions from those they think their contributors will accept, for the purpose of getting them emotionally involved or off dead center.

it. That is why they are engineers in the first place. Without such a device the project engineer, with all his contributors, makes special efforts to keep the team working together toward the desired goal.

This puts even heavier emphasis on some of the control techniques discussed in Chapter 4. Project meetings and memoranda covering them must be especially effective. I have seen large, fairly successful hardware projects headed by an engineer who ran infrequent and miserable group conferences with no summarizing memoranda. He was partly able to make up for this lack by an excellent effort on specifications. But I have never seen a large study that amounted to anything during which reasonably effective project meetings were not held. The contributors have nothing else to go on.

Apparent exceptions to this observation are all examples of one-man efforts wherein the others contributing are doing far less than they are capable of. The total results are either much poorer than they should be or are gained at frightful inefficiency.

One rather odd characteristic of large study control is that the success of some contributing groups carries a penalty for them. Suppose that on a $40,000 dollar study three of the groups participating are budgeted $8000 each. Group A finishes its work (answers the appropriate questions) after spending only half its money and time. This may occur because these engineers are particularly good, or simply because the nature of their problem lent itself to rapid solution, or, more likely, some combination of these possibilities. The reason makes no difference at all. The project engineer, alert to take advantage of the flexibility inherent in this kind of work, cuts off the group's remaining time and funds to use them elsewhere. He may use these funds to extend the work of another group which isn't doing nearly as well but whose answers are badly needed.

On first sight this is rank injustice. If improperly handled, it will result in contributors hiding results and dragging out work. It is primarily a management problem, but everyone should recognize it and help with the solution. To take a less than professional attitude as either a contributor or a group leader is unacceptable.

The flexibility of studies is used in just this kind of way. A study normally ends when the allotted funds are exhausted. (There are important exceptions.) The problem is to do the most that can be done within the funds and time available. It is a matter of balance. To have one link of the project chain well forged but another hardly started is useless. For example, in our tunnel diode investigation we noted that it would be foolish to work out excellent circuit possibilities in detail without inquiring into the temperature sensitivity of the devices.

Unfortunately, the project engineer does not know the proper balance to start with. He makes his first guess in the initial budget breakdown. He knows this estimate will change in almost every project. He is alert to take advantages of breakthroughs and to shore up more difficult areas where desirable. Perhaps entirely new areas for investigation will be revealed in the progress made to date.

It is difficult to redirect an amorphous project in which little direction was established to begin with. Breakthroughs will not be recognized, nor will problem areas. Again you can see the need for a specific overall objective and intermediate goals in study work, even though the latter may be frequently modified.

The flexibility inherent in study work can also be exploited in organizing. Work is effectively organized around people and available competence rather than around hardware components. As long as workable interfaces are created, there are endless structural possibilities for most large studies. The project engineer avoids accepting the constraints of normal engineering organization when it is advantageous to do otherwise.

Study Reports

Just as the somewhat nebulous nature of most studies demands careful project control, so the report demands careful organization. There is no shiny piece of equipment or expensive structure to draw attention from the report or to compensate for its inadequacies.

Answers to the questions that prompted the study should stand out clearly. If there are no answers, that fact should be stated unmistakably. If the answers are in some degree tentative, as they almost always will be, that fact should be made perfectly clear. Include an indication of the amount of confidence justified.

Engineers seem always afraid that their reports will be misinterpreted. A higher boss or an enthusiastic salesman, the engineer thinks, will take his estimates or tentative answers as firm and hold him accountable for them. The best defense against this danger is to *make the answers and the degree of confidence to be placed in them stand out boldly.*

Hiding answers and conclusions in the text or obscuring them by dozens of qualifications will not accomplish the same result. One of the best ways to indicate sureness or the lack of it in a result is to list the critical factors still involved. Then the reader can judge for himself how serious these are. In our tunnel diode preliminary study, for example, if the most critical factor listed as remaining were the question of whether the price of applicable devices would fall much below thirteen cents apiece, this would tell one story. But if the most critical factor were the question of whether a recently announced but currently unavailable device could be brought by its vendor to satisfactory and economical production in the next six months, other conclusions would be drawn.

Good Engineering Practices in Study Work

1. State the problem carefully in terms of the result to be obtained rather than in terms of what is to be studied.
2. Organize project work with particular regard to the expert individuals and groups available to participate.
3. Establish special means for integrating the work of the contributing groups and individuals.
4. Make particular provision to overcome the lack of the hardware stimulus.
5. Bring participants together frequently so that all may continue to appreciate and direct their efforts toward the overall goal, and so that all may assist in correcting divergences between contributing groups.
6. Provide regular status summaries to all participants.
7. Revise overall program and time and cost schedules as needed to take advantage of study breakthroughs. Abandon unfruitful approaches as soon as possible.

Poor Engineering Practices in Study Work

1. Begin the project with the general objective of studying certain areas.
2. Assume that, because of the greater flexibility afforded in study work, scheduling and control are not needed or unimportant.
3. Assume that careful and more or less independent study of each element of the problem will by itself add up to an overall solution.
4. Allow one or two areas of the study to dominate in time, money, and influence simply because of the availability of experienced people in those areas.
5. Allow successful groups to continue their work on originally scheduled funds after additional effort is no longer needed.

Chapter Twelve

Systems

Many ideas from systems work should be applied more widely to all engineering. All of your work can benefit by your understanding them.

For the past ten or fifteen years the term "systems" has been a fashionable one in engineering. The interest in systems grew principally from the war and large-scale defense projects, particularly missiles and aircraft. Large NASA projects such as "Apollo" have continued to stimulate professional interest in the systems concept.

In brief, *systems engineering is putting together large and complex groups of equipment to operate in a specific environment, usually with people, so as to accomplish better some goal.*

Having read the past few chapters, you'll ask, "What's new about that?" That question has been raised by quite a few experienced engineers. Except for its emphasis on the words "large" and "complex" this definition could apply to almost any project. What has happened since the war is that project size and complexity have so increased that earlier integration techniques and practices are no longer adequate. Before going into these things further, let us look at some examples of systems.

If your project involves designing a single machine for some factory, few would call it systems work. But if you were to start with a whole manufacturing process (to be set up in a new building) and determine how the machines and the entire process flow were to be designed, this project could be considered systems work. In addition to designing or selecting the individual machines, you would have to decide how best to use the people available, how to handle and transport the material and parts between machine stations, how to control the entire process, and so forth.

Suppose that a new field piece were being considered for the Army's artillery. If you were called upon to work out an effective procedure for operating the piece and a drill for the gun squad, this would not constitute systems work in the eyes of most people. But if you were asked to work out the entire problem of integrating the new weapon into the artillery, it could be systems work of the first order. You would consider unit organization to use the piece, the tactics appropriate to the weapon, the ammunition supply requirements and how they would affect present supply procedures, the effect of the weapon's characteristics on the artillery support methods for an infantry division, and many more things. And you would do this in the light of environment—the present training and organizational concepts, climatic conditions, and maintenance capabilities, for example.

Designing a single aircraft might be considered systems work by some but not by others. It certainly involves integrating a number of complex subsystems into a useful whole in such a way that a human being can fly it. Usually, however, the design of commercial transport aircraft is thought of in terms of desgning a system involving not only the aircraft itself but also things like crew size and training; basing requirements, including runway length and load-carrying ability; maintenance facilities; signal communication; flight profiles; and turn-around time.

Much thought is being given today to improving mass transportation means in cities. The overall problem here could hardly be regarded in any other light than as a systems problem, including as it does studies of traffic movement (which varies probabilistically from day to day and season to season), geography, political considerations, and technical design alternatives between rail, street, and air.

From these examples we can see that systems work usually involves the following:

> A big project.
> Many diverse parts of the project.
> Complex interrelation between parts.
> Heavy involvement with environment.
> The human element.
> Many trade-offs in design.
> Large probabilistic content.

Systems and Complexity

It is useful to order the elements of an engineer's world on a scale of increasing complexity like this:

> Materials.
> Components.
> Assemblies.
> Subsystems.
> Systems.

You can break this list down further, if you like. Some projects will deal with just one of these ideas, some with two or more. Items earlier in the list are combined to make later items. Here are examples of this kind of ordering from the electronics industry, in the same categories:

> Foil, glass, polymers, germanium.
> Capacitors, transistors.
> Amplifiers, oscillators.
> Radio-transmitting station.
> Broadcasting industry.

We might try another example from the automobile industry. Of course the items listed in these examples are only a few from many possible ones.

Rubber, steel, glass.
Brake shoes, tires.
Trucks, passenger cars.
Automobile dealer.
All the cars, streets, traffic control devices and organization in the city of Worcester.

Observe that all this is relative. There is no agreement as to what constitutes a component or a system. In our automobile industry example a man running a large bus company might consider that his company is the system. Clutches could be assemblies of buses (his subsystems).

It is an old (and not very plausible) joke in the electronics industry that physicists consider resistors to be systems. They are, in a sense, when you consider the complex interrelation of atoms and molecules under the impulsion of electrical stresses in an environment including, for example, moisture. But to a manufacturer who buys resistors from vendors they are in effect materials or, at the most, components.

At any rate the systems concept stands near the end of the complexity scale. *A system is an interconnected complex of functionally related elements.* If you don't like the first definition, try this one.

Expanding Your Engineering Outlook

How limited is your engineering thinking? Consider this story. In the Middle Ages an infant was incarcerated with his mother in a dungeon. The mother soon died, but the boy lived on alone in the dark, damp, and cold, eating the coarsest food and receiving no attention from anyone. Rescued when he was a youth, he desired only to return to his dungeon. He was unaccustomed to warmth, sunlight hurt his eyes, he couldn't eat choice foods. The beautiful outside world and association with others held only terror for him.

In a similar way it appears that we are all prisoners of our past. We are formed by the sum of previous experiences.

This analogy holds true for your engineering thinking, too. You must make opportunities for it to grow and develop. You have no reason to suppose that you can jump suddenly into the sunshine of successful practice. Most engineers start out with responsibility in a small area of design or investigation. Perhaps the new mechanical engineer works on the drive mechanism for a magnetic recorder. If he does well he may later be given responsibility for the entire tapedeck. There is considerably more to be considered in this new work than in the old. Further progress may put him in an entirely new field with interests in more diverse equipments. Perhaps he participates in sales work. As his experiences and interests broaden, he becomes more aware of the total environment in which his engineering work and product must live. He begins to understand the idea of systems.

But we have seen earlier that every engineer must have a wide outlook. How can he be helping to better meet human need if he is unable to appreciate what that need is? How can his designs work in a real world environment if he hasn't fully taken that environment into engineering consideration? How can he know to do this unless he has experienced that environment?

In many ways the requirements of systems engineering are for nothing more than the same kind of excellence which should characterize all our engineering work. We need to emerge from the dungeon of narrow technological thinking and open our eyes to the possibilities which have always been here. In doing so we do not give up our meticulous and expert attention to the details of a specialty, but rather add to these an appreciation of how the specialty fits into everything else.

Organizing a Systems Project

The term "project team" was limited in Chapters 3 and 4 to a single group of engineers and their supporters—perhaps two to six engineers with assisting draftsmen, technicians, and others. Systems projects are usually too big to be done by a single team. Instead the project manager organizes a number of such groups, one to cover each of the areas of work needed. Each group is run by what we have been calling a project engineer.

Figure 12-1 shows a simplified organization example for an assumed study project on bomber defense. Let's assume in this organizational example that you are to examine the defensive possibilities of radar, infrared (IR), armament, and computer control—all in the operational environment of a future tactical situation.

In Chapters 3 and 4 we examined in some detail the relations between members of a project team. We saw in particular how necessary it is for each member to integrate his own work into the total effort with careful attention to interfaces between his contribution and that of his associates. This kind of integration goes on within each of the larger project's team groups shown in the figure.

In addition the groups must integrate work between each other to produce the total system. In principle groups cooperate in solving the total systems problem in the same way that each team solves its individual problems. Project engineers of each team join as a systems team under the leadership of the project manager. In practice things work out a little differently.

We have seen the complexity of interfaces between team members of even a simple project. Naturally, between the groups on a large systems project integration problems are bigger. Even worse, a complete solution for environmental compatibility cannot be reached by the groups alone if their equipments are interrelated functionally.

For example, in considering the defense of a bomber, a major environmental element is the attacks that it will encounter during its mission. The interface between the radar group and the armament group in this environmental situation is so complex that neither could be expected to solve it within the ordinary scope of its specialty. Thwarting these attacks is not really either a radar or an armament problem.

More generally there is a new element in systems work not provided for by the first five groups reading from the left, shown in Fig. 12-1. There will be many constraints (including environmental) placed on

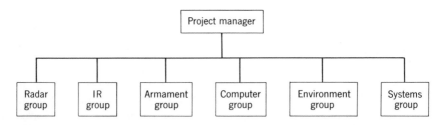

Fig. 12-1. A simplified project organization for a system study. This project manager has set up six teams, each headed by a project engineer, to accomplish the work.

the system as a whole whose effect on the individual subsystems (radar, IR, computer, etc.) is not at all obvious. For example, in the bomber problem there will surely be a practical limit to what an airborne defensive system can weigh, but how should this limit be divided between the various groups? As another example, suppose that a certain hit probability is required from some of the weapons. How shall the inaccuracies and uncertainties be apportioned between the fire control radar and the armament?

Another problem in getting project work accomplished with the first five groups in Fig. 12-1 is that actually not one of them could normally attack the main issue—how to defend a bomber. Each group is ready to apply its *special* competence to the problem. The radar people will determine how best they will and how well they can give long-range surveillance and then fire control. The IR people will be interested in the same kind of things. The armament team will examine which types of armament will be suitable for defense and how effective they can be under given surveillance and control situations. The computer block will see what is required and possible to effect fire control, warning, weapons assignment, and so forth. By means of studies the environmental group will determine the total environment to be encountered in the future period during which the bomber is to fly. But none of these groups is really individually attacking the whole problem for which your project was set up.

For these three reasons (magnitude and complexity of the integration problem, collective nature of constraints, need to attack the overall problem) it is customary to organize a systems team as shown on the right of Fig. 12-1. This systems team, headed by a project engineer, is similar to the other groups in some ways but different in others. Its work and relation to the rest of the project is sometimes misunderstood, even by the systems engineers who man it, with consequent harm to project efficiency.

Let us look at what this systems group does and how its members act.

The Systems Group

Large systems projects need careful integration of all efforts, just as small one-team projects do. It is even less feasible for the project manager to perform all this integration with several dozen or even several hundred professional contributors. Each engineer is still responsible for ascertaining that his work fits into the whole.

Because of the large number of engineers and consequent potential interfaces, working meetings of the whole project are impractical. Each group conducts itself as a project team. You will remember that the project engineers of these teams are responsible for their groups and provide leadership for technical and administrative integration of the work.

In the same way the project manager provides for integration of group efforts. He uses many of the means discussed in Chapter 3. Instead of project meetings of all contributors he holds project meetings of the group leaders (project engineers) or their representatives. We will refer to this type of conference as the *project meeting*.

Like the team meetings on smaller jobs, the project meeting in effect controls the project in that it works out solutions. But these meetings are seldom held as frequently as those within the teams. Also, most of the detailed integration work is not done in the meetings; instead the project meeting lays out this work (and evaluates what has been done) but delegates the actual performance to a systems group. There the system specifications, alternate configurations, constraint breakdowns to groups, analysis of system performance, and analysis of trade-offs across group interfaces are made and remade as the project progresses.

In *one sense* the systems group solves the overall problem of how to design the entire system. Radar, IR, armament, and computer groups provide the design of specific equipments in their fields. It is more useful, however, to consider that the project meetings (assembly of group leaders, including the systems project engineer) solve the whole problem and that the systems group is merely the instrument of performance.

The systems group does its work almost exclusively by paper and computer. It needs constant inputs from hardware groups and others, which in turn require inputs from the systems group. There is a continual round of informal conferences between systems engineers and their hardware colleagues. This is really no different from the interchange between engineers within project teams, but it is more extensive and may be more formal.

You can see that the systems group is vital to a large project. A few large projects may be mere aggregations of more or less independent parts (and therefore do not meet our systems definition very well). With these exceptions it is hard to see how this work can be accomplished by any but a special group, regardless of what it is called.

Sometimes a project manager instead of forming a systems team will retain some systems people on his "staff" and do the work himself. Such an organization works satisfactorily on occasion. Alternatively, a

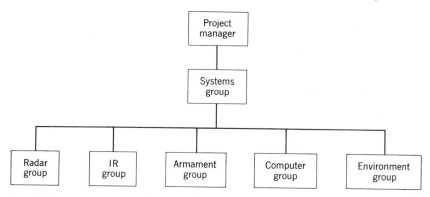

Fig. 12-2. A second type of systems project organization. Here all other technical groups are subordinated to the systems group. This arrangement is sometimes used effectively where there is relatively little interrelation between groups or where relatively little detail innovation is sought. In many cases it results in poorer project effort.

project manager may expect the meetings of project engineers (group leaders) to accomplish this detail work. This is seldom successful, since these men are busy with their group problems.

Figure 12-2 shows another organization wherein the systems group runs the project and all others report to it. This is effective when (a) hardware details are simple and noncritical, or (b) individual project engineers are not very adept at cooperation and teamwork, or (c) the project manager wants to run a very tight and closely controlled job. In this case he no longer uses a project meeting for technical control.

Usually this is a poor organization because it doesn't bring out the best efforts in the detail groups, particularly when critical factors are involved. An experienced radar engineer, for example, is less likely to break through on a critical factor when he is inhibited by junior systems people. If he is put in the systems group, he will miss the stimulation of his colleagues (Chapter 3).

With engineers who have had no systems experience there is a tendency for systems people and hardware people to divide and fight. You can guess the solution to this: (a) each needs to understand the functions of the other, and (b) all need to keep the project goal in mind. In these big projects (as in smaller ones) no parts of the work are more important than others. Systems people are just as vital as hardware people but no more so. Breakthrough ideas come from unexpected sources. All participants contribute to the hundreds of ideas that make up the project result.

Two Approaches to Systems Work

How do those involved in a systems project start on their work?

The Systems Approach

In the systems approach systems engineers first examine the overall problem and decide how it will be worked out. They draw up specific requirements for each hardware group and specify all interface conditions between groups so that no problems will exist there. Second, the hardware groups design their equipments. Thus, one might suppose, the project goal will be accomplished. This approach appeals to quite a few systems engineers.

The Hardware Approach

A different approach is more appealing to many hardware engineers. Each hardware group first designs the best equipment it can to accomplish its part of what it feels the task to be. *Then* the systems engineers take these results and combine them into the solution. In both these approaches the hardware effort is normally the major portion of the work, although it might use 60% of the time and funds in the first approach and 90% in the second.

You might suspect, by analogy to the project team, that neither of these solutions will work except in unusual and largely trivial cases. The first is inadequate because systems people do not know enough about what can be developed in hardware. They cannot properly specify equipments before detailed investigation and development. They either shoot too high and require equipment performance which cannot be attained, or else they are too conservative and settle for far less than could be had. The first mistake torpedoes the project. Competition will quickly uncover the second.

Our second approach is unrealistic because hardware groups do not know to begin with (in any workable detail) what their overall task is. Systems projects are complex enough that individual group missions cannot be simply derived from the overall system mission. And how will the common constraints be apportioned?

The Real Approach

In practice both approaches described are used at the same time. The systems engineers develop initial specifications. The hardware engineers' first output comes usually as state-of-the-art estimates. Heavy communication between the groups starts at once. The first steps of the systems people serve to partially channel and redirect the hardware effort. Hardware's first estimates of attainable performance send the systems engineers back to their block diagrams to revise overall configurations and specifications. So it goes throughout the project.

As time goes on, the work of each group firms, becomes less tentative, embodies more detail. Major decisions are taken with respect to principal alternatives for the whole project and for each hardware area. The work of each group is channeled ever more closely into the final configuration. Methods of systems analysis developed earlier are put to work on the congealing system and refined. As the project draws to a close, a detailed, reasonably optimum system emerges which can be expected with confidence to meet the requirements set for it.

Systems Engineers

Now that we have seen what systems work is, we need to know more about the systems engineer himself. So far it would appear that he is someone who is assigned to a systems group on a large project. Actually the term "systems engineer" can mean two things: first, our man who assigned to the systems group; and, second, one who specializes in the analytical methods peculiar to systems work.

Many good systems engineers come out of hardware specialties. It's easy to see why. In most systems projects strategic factors are uncovered in hardware areas. For example, one is able to build a better defensive system, perhaps, because of a superior radar idea. The concept might be useless without careful system configuration to take advantage of it, but the breakthrough comes in the hardware area (in this example) rather than in systems work. If a systems man has done successful work in some area of hardware, he is able to appreciate what is involved and needed in that kind of effort.

Other systems engineers come from the ranks of those trained in special systems techniques. These techniques are heavily mathematical. We will identify some of them in the next section. Most systems are so complex that it is hard for anyone not acquainted with these sophisticated methods to really understand this work.

If the engineer with a background in some hardware discipline is to be a capable systems man, he will have to overcome his lack of analytical techniques. Conversely, if the mathematician is to be a good engineering systems man, he must learn something about hardware limitations and development work.

Even those hardware-oriented engineers who will continue to grow within these specialties will find that some knowledge of new systems techniques can broaden their appreciation of their own field.

Some Special Systems Tools and Ideas

The analytical methods peculiar to systems work, mentioned above, need not scare anybody. Certainly it is desirable to have a knowledge of at least what they mean and what kind of problem situations they apply to.

As an engineer you will frequently encounter situations that you don't understand; on these occasions you will need more advanced methods than you have already at your command. In digging into these techniques when you need them—or, indeed, to see whether you need them—first consult someone experienced with their use if you can. (But beware of the "expert problem," discussed in Chapter 18.) If necessary, you can still do a lot with textbooks alone.

Here are very brief explanations of a few terms to get you started.

Modeling is the simplified, *approximate representation* of a problem so that mathematical or physical operations on the "model" will show the engineer how the actual situation would develop or change.

For example, it might be desired to know the effect of back seat passengers on automobile performance. They could be modeled simply as 200-pound, stiff, rectangular boxes. The boxes could actually be built and strapped into the back seat of a test car. Or they could be merely postulated for mathematical analysis. This kind of model should be adequate to test, say, the effect on engine performance. For some other investigations, such as the adequacy of internal fittings for passenger safety, a more complex and realistic model would be required. You would probably build it with two or more joints and some realistic distribution of body mass.

The equations which (you assume) will satisfactorily represent your problem are often thought of as a *mathematical model*. Of course, any reasonably interesting engineering problem is so complex that simplifying assumptions are always made. For example, little in this world is actually linear, but we almost always use linear models. They are usually thought to be adequate for engineering purposes in "the region of interest."

Probability and statistics is that branch of mathematics dealing with the analysis of random or chance events and the treatment of large numbers of data. Most engineering or business systems are subject to random processes in one way or another; the larger the systems, the more important this consideration is likely to be. If you were not exposed to a great deal of this kind of thinking in your undergraduate years, take a basic course in the subject or work through a simple text.

Here's a good example of the importance of random events on engineering systems. In a transit system the number of people who will get on or off at a given stop is subject to wide variation. Some of this is partly predictable; some of it is not. Spacing between trains is similarly subject to chance variation. Nevertheless the system as a whole must be designed to operate satisfactorily under these variations.

Allocation problems deal with a limited quantity of some problem element which must be distributed among a number of using or producing subsystems. They are very common (and varied) in business situations and are becoming frequent in more purely technical systems.

Suppose that you are designing the food service facilities for a new airplane but can use only 800 pounds of weight. How will you allocate this weight among the various parts of your subsystem for optimum results?

This example calls attention to an interesting feature of much systems work: the difficulty of deciding what "optimizing" means. Often it's hard to set the *criterion* for optimization, but not here—it's mainly customer satisfaction, although there are some technical constraints. But how in the world are you going to *measure* the effect on "customer satisfaction" of each of the elements of the system—trays and silver, food quality, temperature of hot and cold dishes, speed of service, variety, convenience to hostesses, etc.?

Mathematical programming is one kind of solution of allocation problems and is mostly limited to *linear* programming. *Dynamic* programming includes problems some of whose parameters change, usually with time. A typical programming problem would be to determine the proper allocation of a given batch of orders among variously located plants.

Queuing (or waiting line) problems and *sequencing* (or ordering) problems consider the design of systems wherein some operational elements queue-up or wait to be processed. For example, how frequently should buses be run on a given line? If they are too far apart, there will be long waits and dissatisfaction; if too close together, bus size or loading may be uneconomic. As another example, in a maintenance shop how much priority should be given to what kinds of jobs?

Operations research (OR) is a general term including the development and use of all the techniques discussed above and many others. In one sense it is synonymous with systems engineering. The concept sprang from successful mathematical (mostly probability) analysis of tactical operations during World War II and has been applied extensively to business since.

Perhaps OR is usually distinguishable from systems engineering in that most OR people operate as consultants to fairly high levels of management. People trained in OR are often invaluable members of the systems group on engineering projects, as discussed earlier in this chapter.

Game theory applies the concepts and techniques above (and others) to situations involving conflicting decision-making systems, as in business competition or war.

Much of the special technique of systems analysis (or operations research) is in a poor or nongeneral degree of advance. It is of little use at present to the engineer. Game theory is an example of such a tool. Although it provides some clarifying concepts, it offers little of use for analytical purpose at present.

Simulation is the use of models, as mentioned above, for "make believe" operation of systems to see how they perform. It can be done with physical models but is especially valuable with mathematical models on high-speed computers, where time compression permits long-range or multiple trials during system design work.

The *Monte Carlo* method is a form of simulation in which system inputs are randomly generated, perhaps by computer. You might, for example, "run" an airline for a few days on a computer by taking randomly generated numbers of passengers at each airport. Such an investigation could give you a good idea of the loading of scheduled flights and passenger queuing at terminals.

Many of these concepts and methods have produced startling improvements in business organization or system design and operation. Others are as yet hardly more than academic toys. Much sound investigation is continuing in these areas, and more information will be available for the engineer in the future.

The technical jargon usual in a new area has mushroomed to a ludicrous degree in systems work. Some enthusiasts seem scarcely able to converse with more literal-minded engineers. I once had a boss tell me (referring to statements of an operations research man on a systems project), "I think he's right, but I'll admit that I can't understand what he's talking about." Here, as elsewhere, the competent specialist will understand engineering well enough to be able to explain his specialty and its application to his associates.

Good Practices in Systems Engineering Work

1. Engineers look on the system as a whole, rather than that part of the hardware they happen to be assigned to, as their ultimate goal.

2. Project teams are organized so that the systems people are on the same level as the hardware groups (usually advantageous).
3. Hardware specialists learn enough about systems engineering problems and techniques to be able to operate effectively on a project team working on part of a large system.
4. Engineers recognize that the work is conducted by a continual series of hardware systems trade-offs. The system planning becomes more realistic with respect to what can actually be accomplished hardware-wise, and the hardware progressively approaches a reasonable optimum for systems purposes.

Poor Practices in Systems Engineering Work

1. Systems people and hardware people continue in such ignorance of each other's function that consequent misunderstandings and hostility harm the work.
2. Systems engineers first write more or less firm specifications for hardware groups, which must be met to complete the project.
3. It is assumed that there is nothing more to systems work than hooking up hardware when it has been designed.
4. The systems project is organized so that hardware groups are technically subordinate to the systems group (usually a mistake).

Chapter Thirteen

Proposal Work

In almost any organization you will frequently be involved in proposal work. These intensive short projects can be real opportunities for the group and for you. They are among the most interesting and stimulating activities of engineering. The relationship of a proposal to a long engineering project is about that of a raid to a set military engagement.

A proposal is usually pretty hectic. You work all night on some proposals. Engineers are pulled temporarily from other jobs to fill in. Information is scarce to nonexistent, but there is much room for judgment and imagination. Like a battle, it brings out the best in most participants, spotlighting those who can think on their feet.

Can you keep a sense of direction and purpose in the midst of confusion? Frequently during proposal work some young engineer previously buried in a large organization will step to the front with real contributions and leadership, never to lapse into obscurity again.

Proposals are, of course, offers to do business along certain lines. As in most other human activity, they are a prominent part of engineering. If you have a good idea to improve efficiency in your organization and take it to your boss to sell him on letting you try it, that's a simple form of proposal. If the Air Force announces that it will receive technical ideas and bids to accomplish certain engineering work, your company's response to this invitation is a proposal. Proposals can be written or oral and often are both. They run the gamut of complexity from a simple oral request, requiring 30 minutes' preparation of the convincing, supporting ideas, to twenty-volume, six-month, $500,000 engineering efforts.

The One (and Only) Purpose of Proposals

Right at the outset you will want to recognize that a proposal has one primary purpose—*to sell something*. It is a presentation, oral or written, undertaken for the purpose of getting another party—person, company, group—to do something you want it to. Usually it will be to employ your company's services in some capacity for development, design, construction, or some other purpose.

You design the presentation to sell. *All* preparatory work on a proposal project is aimed at coming up with a selling presentation. There will be more on this later.

I hope that the idea of selling something doesn't surprise you. The engineer has to continually sell ideas to his boss. We have all seen people with good ideas who were passed over simply because they didn't convince anyone.

Traditionally, the engineer had to sell himself to an employer at the beginning of his active career. Maybe in this day of so-called engineer shortage this situation is hard to visualize. Similarly, the employer needed to sell himself to the engineer.* But selling in both directions has always gone on and always will. Even earlier, as a student you sold your solution ideas and projects to professors.

But proposal work has a special significance in engineering. Physical science (as well as most of technology) does not have a selling appeal in itself. Human needs are met directly in terms of food, clothing, shelter, and basic emotional ideas. Thus you have to sell your services in terms of human need. That is what a selling proposal presentation does. Even when you are responding to an invitation to bid on highly technical development, for instance, you are selling to *people* and cannot overlook this fact.

Your boss with his rusty old M.E. degree won't buy your solution just because it has a Laplace transform in it or an ingenious circuit. He'll buy to satisfy his needs. His need is more likely to be confidence that your project will be completed on time. Or he may need something to get around a new technical breakthrough by a competitor. Furthermore, important as it is that we engineers understand and correctly use Laplace transforms, you'll have to admit that they will satisfy very few people's emotional needs.

Hence an engineer frequently prepares proposals—sometimes formally, more often informally. Formal proposal work is usually handled

---

* There is a shortage of good engineers and probably always will be, but it is hard to believe that the present demand for mere numbers is realistic. A few excellent engineers operating professionally, supported by other technical help, can accomplish more than a large number of moderately good engineers. I have seen projects where *reducing* the number of "engineers" by cutting off some of the dead wood actually *increased* productivity. People who do not operate effectively and imaginatively as part of a project team create so many integration and communication problems that their contributions cannot compensate for the damage they do.

as a short study project. Most requirements and paraphernalia for project work hold here too. For instance, a proposal or project engineer leads the work, doing a fast but careful job of scheduling and project integration.

We will keep in mind, however, that the objective of proposal work is quite different from that of the usual project.

You May Get the Job

Engineering organizations propose work that they would like to do. They live, in effect, on that fraction of their proposals that actually is translated into action—the proposals they have made which the customer likes. Therefore it is startling on occasion to hear an engineering manager, reviewing his list of outstanding proposals, say, "I'm afraid we're liable to get that one." What he means, of course, is that his group proposed and bid on some very difficult job.

Since the hectic and enthusiastic proposal-writing days, he has come to wonder whether he can really perform on the job as proposed. So don't lose sight of the fact that you may well be awarded the jobs on which you bid. You don't want to promise so much and bid so low that you can't come through with performance when the time comes.

To this end a major part of proposal work is configuring a feasible device or structure or system to do the job. Something has to be worked out far enough to be sure (or reasonably sure) that it can be carried out if the job comes in. If your customer is smart—and most of them are—he'll look in your proposal for evidence that you have established the feasibility of what you propose, at least reasonably well. Also, in most instances he'll want to be convinced that your costing was done reasonably too. All this adds to the salability and plausibility of your proposal.

Needless to say, the slightest dishonesty is no more acceptable in proposal work than in any other phase of engineering. It is incompatible with your mission to meet human need.

Sometimes a person may be taken in by the ridiculous doctrine that the end justifies the means, that good ends can be accomplished by evil means. Such thinking denies the orderly and consistent reality on which your technology rests. Risks involved in a proposed undertaking must be *clearly shown to the prospective customer*. Risks involved in proposed time schedules and (unless the price is fixed) in costing should be shared also.

Thus all proposal work still has selling as its sole purpose. Even though you must do enough feasibility work to be sure you have something to sell, this constitutes part of the honest sales effort.

Quite a few engineers just don't understand this fundamental point about proposal work—that *all* effort must be tailored to and aimed at the selling presentation itself. This is apparently part of the misconception which makes people think of technical work as some sort of paid hobby. They think all technically inclined young men experiment in laboratory work or design for the sheer joy of it. It is all fun—if you and I didn't think so, we would have no business being engineers. But we know what the real mission of engineering is. Put your effort directly toward this goal; don't dissipate it elsewhere.

Control Is Easy

Control of a proposal project although not difficult, unfortunately is frequently slighted. There is a very limited amount of time with almost no possibility of extending it. Time is so short, in fact, that the usual ironing out of difficulties and inconsistencies that comes about naturally through cross-communication hardly has time to occur. People are assigned temporarily on an *ad hoc* basis that would seem to work against inspiration and team effort. Technical information is scarce on large parts of the work, and there is little time to dig it out.

Actually these difficulties are easily overcome because of two special elements in proposal work. First, the very vagueness of the situation and the haste required create an environment conducive to creativity and dedicated effort. When the chips are down and the organization in dire need, what man worthy of his calling could help but pitch in enthusiastically? If there is no time to find out every detail of earlier solutions, you've *got* to invent.

In the second place, control is relatively easy because the goal is immediate and clear. A project engineer who understands these points can easily motivate and control his proposal team's effort in terms of the goal—selling the job to the customer.

Use the Book

To ease the situation even more, you can make this goal so concrete that no one can miss it. We saw earlier that the selling proposal could be written or oral, but for a proposal of any size there will certainly be a book. Thus we can take the book itself as the specific goal. Using the book as a focus for all engineering effort is a unifying and controlling tool of real power in conducting this kind of work.

A common misapprehension is that engineers spend perhaps 90% of proposal time in investigation and ideation. Then in the remaining time everyone sits down together to write a book. This viewpoint throws away the integrating power of the book as a control tool. It will result in poorer, wasteful effort. It will inevitably result in a poorer selling presentation.

Instead the book is planned from the beginning. At the start of the proposal effort a tentative outline or table of contents is given to all participants. As work progresses this outline is modified and filled out. Very early, too, writing responsibilities are assigned.

A Technical Approach?

Another common misapprehension about proposal books is that they are written first (by engineers) from a technical point of view. Then marketing dresses them up with a preface and new conclusions to make them selling tools.

Nothing could be further from the proper organization of presentation material. The book is organized from the beginning as a selling document. No doubt, since it pertains to an engineering problem, a great deal of technology will be included, and rightfully so, but the purpose of this content is solely to "beef up" the sell. Now remember that we use the idea of selling in its best and highest sense.* Selling is showing someone how his needs can be met effectively and economically. You meet these needs by the application of technology, but this is not technology for technology's sake.

As an example of this difference in approach suppose that the design and production of a specific kind of machine was being proposed. A purely technical approach would probably be to describe the equipment part by part, discuss the function of each part, and then observe that such equipment would do the job the customer wanted to have done.

Although all that information is probably needed somewhere in the proposal, there is a more effective selling approach. First discuss the customer's principal requirements and concerns about his application. Then show how the proposed equipment will meet each of the requirements and overcome each of the concerns. Go on to show why other equipments designed in the obvious alternative ways could not do anywhere near as good a job. Include enough honest justification so that the customer can be convinced your claims are valid.

Let us summarize all this: organize the proposal book and sell around

---

* There is more on selling in Chapter 16.

your customer and his needs, rather than around yourself and your ideas. Your mission is to solve his problem, not to peddle a certain piece of equipment. Incidentally, the customer-oriented approach is the most effective way to sell your product, as many a wise, prosperous salesman has discovered.

Keep It Simple

Excellent engineering is almost invariably simple engineering. When your design and schemes become complicated, watch out!

In proposal work you have a wonderful opportunity for real creativity. There are many different ways to do things. Some are simple and some complex. When you have a choice, try to make the simple answer work. Very often it is possible to propose a simple system to fill or almost fill your customer's stated requirements. There may be opportunity to propose as an alternate *an addition* to further extend the simple system's capabilities. Usually this is better than proposing a complex, difficult solution to start with. Complex engineering results in equipment which is less reliable, harder to maintain, more difficult to train operators for, usually less flexible, and more costly.

Certain military missile systems provide an example of the feasibility and advantages of simplicity. Assume that a requirement exists for 90% probability of kill against certain targets. The engineer finds that he has a choice. He can meet the requirement in a single missile with an expensive, complicated system design. Or he can build a simpler, less expensive system in which each missile has a 75% kill probability against the required target. For many applications the latter system will be superior. It is only necessary to fire two 75% capability missiles to obtain an overall 90% capability.

Costing

Costing in proposals is closely related to complexity or the lack of it. It is certainly a major factor in proposal success. The basic idea, of course, is to cost the individual elements of the proposed system and then add these figures up. But a project engineer must be careful here.

Each group contributing work and cost estimates will tend to be conservative to allow for unexpected problems. When all these estimates are first totaled, the result is astounding! I have never seen a project of any size where the sum of first estimates was reasonable. To avoid this trouble a project engineer makes his own estimates first for each group. Then he is in a better position to work out a more reasonable cost with them. If he can reassure them that he understands the risks involved, it is easier for him to get realistic figures without a lot of padding.*

Your company's marketing function will usually take the lead in presenting costs to the customer. Often one part of a customer's organization will consider the proposal from a technical point of view and leave cost comparisons to a financial or accounting group. Surprisingly there is little communication between these two groups in some cases, their analyses going to yet another group for final decision.

Financial people sometimes have little or no comprehension of technical things, even to the extent of not understanding the implications of various alternative proposals and combinations. Therefore it is usually best to clearly cost the *simplest combination* of elements that will do the customer's job. Show that additional optional alternatives can be added if desired. It is a rare customer who will subtract items from a quotation in order to compare it fairly with others. Of course, it must be perfectly clear what is being offered and costed.

Proposal Cost

Another kind of cost that is difficult to control is that of the proposal work itself. Because of its hasty and *ad hoc* nature a proposal project can easily consume several times the money originally planned for it. Marketing people, wanting to sell it, will insist on adding more and more effort. As engineering ideas unfold, new directions for investigation will appear. The techniques of project control discussed in Chapter 4 will help here.

It is particularly necessary for a project engineer to have a clear initial understanding with his manager about the resources to be used. Set down this understanding in a memorandum and keep it current. As changes are contemplated, they can be considered in the light of what has already been authorized and planned.

Good Engineering Practices in Proposal Work

1. Make clear and specific decision at the outset on what proposal effort will be made and how much will be spent.

---

* See the discussion of PERT in Chapter 4.

2. Start proposal work by tentatively planning the presentation document.
3. When the inevitable changes in the planned effort are made, make them deliberately and with reference to their effect on original plans.
4. Balance the effort on the proposal in such a way that all aspects and particularly all critical areas are covered reasonably well.
5. Make one person responsible for the entire proposal, and on a full-time basis if possible.
6. Keep substantially everyone on a small proposal aware, on a daily or more frequent basis, of all major activities. On a larger effort, at least the principal leaders must be so informed.
7. For all contributing efforts, prepare a brief written statement or outline of what is expected and when.
8. Include the supervisors of persons contributing from other areas in the distribution of work statements and other general proposal documents.

Poor Engineering Practices in Proposal Work

1. Consider proposal and estimating work to be an unfortunate distraction from true engineering effort.
2. Consider that because of the brief and rushed nature of the work there is no time for organizing the effort.
3. Allow the proposal effort to grow by itself without regard to specific plans.
4. Within the short time available, permit technical contributors to work more or less independently of each other, expecting to integrate their material at the end of the project.
5. Accomplish integration largely on a bilateral basis between the project engineer and each individual contributor.
6. Draft the proposal book primarily from a technical standpoint, expecting to add a few marketing paragraphs at the beginning and end.
7. In an effort to excel competition, make the technical proposal so complex that cost is far beyond customer resources and unreliability destroys its usefulness.
8. Cost the work proposed solely from the bottom up, simply totaling the individual estimates of each contributing group.

Chapter Fourteen

The Project Engineer

Every engineer looks forward to the time when he can have a project of his own.

A project engineer has the best job in the business. He has ultimate responsibility for the work as a whole. He is the real architect of the project solution. Even more than his colleagues, he looks at the job as a whole from the beginning. He watches carefully to make all details come together into a timely, economical, fresh, and effective meeting of the need.

The team leader can easily take an independent professional viewpoint. In fact, he must!

You will remember from Chapter 3 that the project engineer is the leader of a team of engineers and supporting people who are assigned a specific project—a problem to solve, a machine to design. We also saw that there are still many one-man projects in engineering. In such a case that man is a project engineer, too. But in this chapter we will think mostly about the team leader.

Often an engineer is first selected for this work because he has made some unusually fine technical contributions to an earlier job. This technical competence and initiative is the most important quality that he brings to the new task, but it is by no means the only one needed. And it *cannot make up for a lack of administrative skill or of leadership.*

It is especially important for an engineer just stepping into his first job as team leader to review the administrative and leadership techniques required. In a well-managed organization his boss will help him with this, but the responsibility is entirely his own. Almost all engineers who come to grief on this first (and possibly last) assignment fail because of poor administration and leadership rather than unsound technical judgment.

Actually the results of a bungled project are so confused that it is usually impossible to separate administrative and technical errors. The excellent project engineer recognizes that his success depends on both the technical and administrative aspects of his work.

Where a Project Engineer Fits in

There are many different ways to organize a business. The blocks of Fig. 14-1 are typical, and we can use them to illustrate some important points in project engineering.

The business is first divided functionally into marketing, engineering, and manufacturing. In addition, there are other functions not quite so large, such as finance, employee and community relations, and legal, some of which are also indicated in the figure. Depending on the size of the business, each of these major functions (we will call them sections) may be further subdivided, but in our chart we have shown only the engineering subdivision.*

A manager of engineering is often called the chief engineer, or sometimes the manager of product design or product conception. Whatever his title, he is the man who is looked to principally for the design of the company's product and in most cases for new technical ideas and inspirations. He also is the chief advisor on technical matters. He doesn't do all these things himself but relies on his organization to support him.

Organizing an Engineering Section

There are two kinds of subdivision under the manager of engineering. First, specific types of *product work* are often separated out—for example, radar engineering or vacuum cleaner engineering. In this way design and conceptual work on certain products can be concentrated in a group which can build up expertness in most of the skills required.

A second kind of subdivision in engineering is *functional*. There may be a circuits engineering group or a mechanical engineering group. There is usually a separate engineering administration subsection. Sometimes all draftsmen are organized into a separate group which can be supported and administered efficiently.

Both kinds of subdividing exist at the same time in most engineering sections. Project engineers are familiar with the details of their company

---

* Designations of the subdivisions of a business at various levels are not at all standardized. Little inference can be drawn from the subnames for any particular concern. One large equipment manufacturer divides his several-billion-dollars-a-year business into groups, divisions, departments (the first level where our Fig. 14-1 functional description clearly applies), sections, subsections, and units. In other companies the term "unit" represents a major subdivision of the whole business.

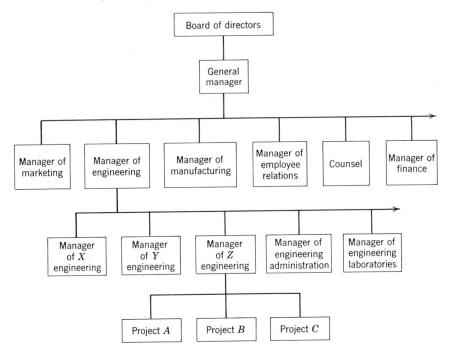

Fig. 14-1. A typical organization for businesses in which there is a large engineering content. Although there is one so-called "Engineering" section, there will be actual engineering work in other sections also, particularly manufacturing and marketing.

organization. They can coordinate their efforts with whatever part of the structure they must go to for support. You may, for example, find yourself working with the engineering administration subsection on matters of standardization. You will have to go to drafting for engineering drawing support.

Often advanced development work is separated clearly by organizing a so-called engineering laboratory. Developers are shielded to some extent from day-to-day distractions of design and production. It is important, however, that such shielding does not separate them from the real world in which their results must stand or fall. In the long run, they must receive their inspiration from that world. A large laboratory will itself have quite an organization structure, possibly similar to the engineering section of Fig. 14-1, including projects and project engineers.

Here we are principally interested in the first product subdivisions, labeled on our chart as X engineering, Y engineering, and Z engineering. Each of these unit managers will have one or more projects going in his group. We have shown the Z area with three projects, each of which is run by a project engineer.

Remember that organization in any company may vary widely from this one and is tailored to particular needs at a specific time. As projects and the nature of the business evolve and change, structure must change to accommodate them. Naturally it is at the lower echelons of Fig. 14-1 that most modifications will take place. Changes at the section level will be slower and less frequent.

In a small company the structure shown under engineering may not be needed at all. It is possible for project teams to report directly to the manager of engineering himself. Larger companies may insert additional levels into the structure.

The Unit Manager

To return to our main topic, we are interested in distinguishing carefully between the work of the unit manager and that of our project engineer. The manager's function is to get the engineering work in his assigned area of responsibility done effectively on a *continuing* basis. He has to think not only about today's jobs—those three projects shown under Z in Fig. 14-1—but about future jobs also. He supplies people and other resources needed for the work. He provides work for his people and meets a budget. He hires and fires his people, trains them, appraises them, and adjusts their salaries. He forms projects and project teams and provides them with support and guidance.

The unit manager must keep himself and his people abreast of the state of the art not only as it changes but also as the company's interests and needs change. A major part of his function is integrating the work and planning of his group into that of the whole company. He keeps his boss informed on (a) the progress and status of his jobs and proposals and (b) the impact of today's state of the art and likely future extensions of it on the company's business.

The Project Engineer

Now contrast all that with the work of a project engineer. His range of work is much more limited. He is responsible for getting his assigned project accomplished on time and within the budget, and with excellent technical results.

The project engineer is a leader of his men—but almost exclusively *from the standpoint of that one project.* He is not responsible for the future of either his engineering area or his men. He is not responsible for their training or progress or appraisals. His efforts are devoted almost entirely to getting the project done through (a) his administrative ability and team leadership and (b) his own outstanding technical understanding and efforts.

The picture is not much different in a large systems project in which several teams participate under a project manager (or even under several layers of project management).° Here continuity is a shorter-range consideration but is nevertheless significant. The unit manager's need for longer-range planning is replaced by the project manager's interest in higher-level technical integration. We have already seen that the project manager will have systems people or a systems group to help him with this goal. His relation to individual engineers, however, is pretty much the same as the unit manager's.

It is important that the project engineer not lose sight of the fact that he also works for the unit manager and must be responsive to his guidance and requirements. Your leadership of a project team cannot conflict significantly with the longer-range goals of your manager. You will be called on to assist the manager with appraisals. No one knows better than you what quality of performance your engineers have given and to a considerable extent what their development needs are, but you are not responsible for the details of this administrative work.

As a leader your effect on young engineers is incalculable. They learn by doing, and hence by performing the tasks you set for them. Their development to project engineer is dependent on what they observe in your own performance. Your successful handling of these secondary aspects of your job, in addition to the successful prosecution of projects themselves, can bring you the opportunity for management responsibility.

Practical Relations with Your Manager

It's an old tactic in military operations to harass the enemy by attacking or threatening to attack his rear. The soldier fights approximately on a linear front toward a given direction with his supplies and support coming from behind him. Since he instinctively realizes he is vulnerable in the rear, he can't fight his part of the battle in front with courage and energy when he is worried about what will happen behind him.

We have a double parallel in the project engineer and his manager. In their relation to each other they stand back to back. Our project engineer directs his effort downward into the project. He is intelligently fighting and overcoming and using nature to solve his problems. His

° See Chapter 12.

unit manager is his source of support and supply. The project engineer goes to him for help when problems seem too large to handle. Especially he relies on the unit manager for continuity, *placing confidence in him to have the next job waiting when this one is completed.* No one can do his best work if he must worry about his boss.

In a similar way the unit manager, as we have seen above, is directing a great deal of his effort away from today's projects to planning for the future, the longer-range unit problems, and to integrating his unit's work effectively with company needs. Though completely responsible for the success of the projects in his unit, he delegates their detailed conduct to his project engineers. In one sense they constitute his rear, his supply and support, since his future and long-range plans are completely dependent on today's successes. *As long as his projects are going well, he can move with confidence into future planning and dealing with the other elements in his company's structure.* If he has to worry about the projects, he can no longer do his best.

Have a Clear Understanding

Now the moral of all this is, of course, that the project engineer and his manager must have a clear understanding with each other at the beginning of the project and then maintain it. Areas of possible misunderstanding are cleared up at the outset or whenever they arise. Since the manager has several projects to consider (as well as other work), it is up to the project engineer to establish this rapport and see that it is uninterrupted. Thus, in military language, you secure your rear and can then energetically and confidently accomplish your project.

Two useful tools in maintaining mutual confidence between manager and project engineer are (a) an initial write-up on how the project will be conducted, and (b) periodic progress reports, usually weekly.

The manager may have his own ideas, possibly somewhat undetailed and vague, about resources to be used in the way of men, money, and facilities, and about the general scheme of attack. The full-time project engineer will have more detailed knowledge and may differ on these points. Thus at the outset he should set down in outline his plan of attack and particularly *the resources that he understands will be given him.* Such a write-up can resolve misunderstanding and prevent future problems. If variations are later made, both parties will understand that they are variations.

Often a job is undertaken after a detailed proposal has been drawn up. Seldom is the proposal accepted as a whole. Thus there may be misunderstandings on how closely the proposal will be followed. These points should be resolved between the manager and his project engineer. Verbal understandings should be confirmed with a *memorandum* by the project engineer as a matter of course. The manager has more on his mind than one project.

Keep Your Manager Informed

Nothing is more worrisome to a manager than a growing doubt about one of his projects. Such misgivings flourish on lack of information. The manager is entitled to have (a) a reasonably up-to-date knowledge of the project progress and problems, and (b) confidence that any really serious development will be brought to his attention immediately.

Often a well-organized manager will set aside a certain hour each week for a more or less formal briefing by his project engineer. This will be in addition to frequent informal visits to project work areas.

It is usually desirable for the project engineer to work up a simple weekly written summary of project status to include technical progress, problems, funds, and time. Sometimes this can be appended to a regular project meeting summary; sometimes it is better left by itself. It is usually distributed to the project even if prepared for the unit manager. When necessary, an additional sheet for the manager only can be included to cover special information—for example, personnel data. See Fig. 14-2.

It is such an encouragement to a project engineer and his team to have the manager fully informed and wholeheartedly behind them that you will wonder why some project engineers resist these simple steps. The fact is that young project engineers sometimes don't understand the relation between technical and administrative work. Also, there are still some engineers who have never overcome a distaste for putting pencil to paper. A few sincere engineers have explained to me that if they spent the time to write these simple reports they wouldn't get the project work done on time. The reply to this objection is that, if you, Mr. Project Engineer, don't know the status of your project well enough to be able to set it down on paper in a few minutes, you had better spend whatever time is necessary to find out.

Some Project-Leading Techniques

Keep First Things First

We noted above that you were probably chosen for your first project engineering job because of technical excellence on related work. You realize this fact and know that on this small project you will be expected

Date: _____   Copies:

      ACTIVITY SUMMARY _____   _____

      PERIOD_____   _____

      BASIS _____   _____

I.   This period--Principal Activities, Successes, Problems:

II.   Next Period--Plans, Events:

III. Financial and Schedule status (estimate as needed):

      Authorized $ _____ Committed and spent $ _____ Left $ _____

      Additional expected $ _____ How? _____

      Schedule status: plus/minus _____ Days/weeks

      COMMENTS:

                            REPORTED BY _____

Fig. 14-2. A simple form for the project engineer's weekly report to his manager.

to perform heavily in actual design and problem solving as well as in leading the project. Funds available will not permit you to sit back without contributing to the details. You cannot afford not to work.

But never lose sight of the fact that your first job is coordination and leading. This is your most important responsibility. Those parts of the project work that you decide to schedule to yourself must not be allowed to interfere with your primary mission—getting the whole job done on schedule, within the budget, and to the specifications.

Allocate Effectively

Project leading is managing the project work; look on it as allocating resources to best meet requirements. The resources at your disposal are people, funds, and time. Your requirements change as progress overcomes difficulties at one point or encounters them some place else. Hence your allocation of resources will vary, but with smoothness. Take into account individual differences in your people's experience and interests to optimize their output.

Keep Details in Place

It is generally better to do a few things well than to scatter resources over many areas. One procedure that I have found useful on projects is this: where there are a number of small, more or less unrelated tasks that arise from time to time but still must be done, limit them to one or two people. Perhaps the project engineer himself can take care of most of them. Thus heavy contributors to critical areas of project work can be left undistracted by other tasks. Wherever possible allow a man to concentrate his full time on one job or at least on one area.

Handle Consultants Carefully

Part-time personnel and consultants are frequently difficult to handle. They cannot integrate as effectively with the project team and are less involved with successful overall results. It is usually best to recognize this limitation and severely restrict the responsibilities entrusted to them. Utilize their efforts as contributions for some full-time team member to integrate into the solution. Occasionally, however, especially when these people work frequently with the same engineers, they can be integrated into the team effort like anyone else.

There is a simple but effective motivating technique especially useful for proposals. Some of your engineers will be lent to you from other sections, perhaps pulled off other urgent work temporarily. Since a man can't do his best if he has doubts about how his boss feels, include each engineer's supervisor on the distribution list for project memoranda. As a bonus you may also get some very knowledgeable help and suggestions free.

Maintain a Division of Labor

It is well known that division of labor was a turning point in the industrial revolution. Instead of everyone doing everything, workers specialize. Thus more can be accomplished with greater efficiency. As a project engineer, make a point of keeping the big picture before all your engineers, but be careful to limit the degree to which you involve your team in problems. Do not destroy the useful concept of a division of labor. On some poorly run projects every member seems to be involved in some way in every task.

Start Your Control Immediately

The project engineer provides for positive control of his project right from the beginning. It is much easier to keep a project under control than to rescue it once it has gotten out of control.

A common mistake is to allow the busyness and confusion of organizing and starting a project to delay time and money scheduling and reporting. Lack of detailed data is sometimes used as a reason for putting off this essential work. However, even if these data are not available, such broad time and money scheduling as can be done should be undertaken. Scheduling and planning usually proceed from the general to the particular and detailed, anyway.

Keep Your Control Effort Current

In large organizations project engineers may rely on the financial or accounting service to provide needed performance data. Often these reports run a week or more behind actual expenditures. During most of the project these reports, combined with your estimates of the past week's expenditures, may be adequate. At critical times (for example, the last few weeks of a project), however, this information is not good enough, and the project will have to keep its own records or arrange to be supplied with current data.

Nothing can relieve or excuse a project leader from the necessity of *knowing* the financial, schedule, and technical status of his work all the time. How can he control his project otherwise?

Stop the Project When It Is Over

We have already noted the difficulty of stopping an engineering project. It should be a major concern of the leader to plan for the end of the work and actually terminate it himself in cooperation with his unit manager.

Leadership and Delegation

As a new project engineer you will want to avoid this mistake: since you were an outstanding contributor to earlier projects and therefore were chosen to lead this one, it will be a temptation to take the work done by your project members, find it unsatisfactory (it's not the way you would have done it), and do it over again yourself. But you must learn to lead rather than to redo.

If you could do the project yourself there would be no need of assigning others to assist you. Also, the effect on the morale of the individual whose work is redone by someone else is quite shattering.

When a team leader has to do any significant amount of rework on his group's output, the fault is his own. He has not managed to communicate to the individual contributor what he wants him to do. This communication is not easy on a complex technical job.

We saw in Chapter 3 that specifications are written on the technical work of each contributor as rapidly as possible. Progress modifies and details them as the work advances. Although the project engineer gets the individual engineer to do this for himself as far as he can, it is the leader's responsibility to review these statements and to see that they are adequate. These specifications start originally as work statements.

The project leader ensures that each engineer's work is effectively guided into the right channels through work statements, through participating in group meetings, and by means of the memoranda of meetings. In addition he sits down with his engineers individually to discuss each one's part. He spends a great deal more time with his beginners than with his veterans, and more time in those areas which are currently critical to the project. Guided thus, the competent individual contributor cannot help but come up with generally acceptable work.

The term "delegation" is often misunderstood. A project engineer delegates certain parts of the work to his associates. He himself is originally responsible for getting them done. After he delegates the work, he is still responsible. He cannot delegate responsibility. He makes the individual contributor responsible to him for the job, but he himself is still responsible to his manager as before. He can't sit back and blame a bungled job on a subordinate.

In effective delegation the leader watches and assists as needed. His help and coaching will vary with need—more to the beginner, less to the veteran. On the other hand he avoids diluting the responsibility and authority he has given by too much help or by unnecessary harassment.

A good rule is to hand out a job assignment and then watch quietly without interference (through the project meetings, for instance) for results. It is better for a man to find his own difficulties and to resolve them if he can. But the leader steps in to help—not take over—when definite trouble appears. In this way new engineers develop, and the leader has more time for his own work.

We have already (in Chapter 5) mentioned the useful technique of a project leader writing the main body of his report himself. This allows more effective delegation of report activities and avoids a great deal of rework.

There is no substitute for leadership from a leader. Although a project engineer is careful to avoid suppressing the original ideas of his group with his own strongly expressed preferences, there comes a time when decisions must be made. After the ideas of the group have been brought out fully, compared, and analyzed, it will often be obscure which is the best course to take. The project engineer naturally makes or approves a decision without delay so that the work can continue.

Hesitancy and vacillation gain nothing but a lowering of team morale. If the leader is developing good technical judgment, most of his decisions will be correct. If he isn't, nothing can help him (certainly not delay). Where an occasional decision is incorrect, this will usually be discovered with least delay and expense by aggressively pursuing the decision as if it were justified. The leader can then reverse himself, when sure of the facts, as gracefully as possible.

A more important aspect of project leadership is creativity. I have seen project engineers who set up a project and monitored it thoroughly without ever making any radical turns or changes. This usually results in mediocre output at best.

Most of the leader's associates are quite well buried in their assigned areas of work. It falls to the leader himself to continually re-examine the work as a whole and to ask whether in the light of new developments some novel departure is in order. Often in project meetings ideas emerge which he alone is in a position to explore. Used creatively, they may lead to the strategic factor for project success.

When you are appointed a project leader, LEAD!

Good Practices for the Project Engineer

1. The unit manager with overall responsibility for the project assigns a project engineer; together they work out the manpower and other resource and support requirements.
2. The project engineer sets down on paper, with manager's approval, the specific project goals, and the resources to be used, including funds and time.
3. The project engineer operates the project in such a way that essentially all professional contributors are aware of every important development or problem on a daily basis.
4. The project engineer keeps himself current on all aspects of his project so that he is ready at any time to explain where the project is technically, financially, and schedule-wise.
5. The project engineer uses to the maximum possible extent the resources and support available to him and his people.
6. The project engineer acts as principal writer and editor of major project reports.

Poor Practices for the Project Engineer

1. The unit manager assigns no project engineer or assigns one on an inadequate part-time basis.
2. The unit manager assigns a "technical" project leader with no responsibility for project administration.
3. The project engineer buries himself in one technical phase of the project to the detriment of other necessary aspects and of his overall integrating role.
4. The unit manager assigns a nominal project engineer but then runs the project himself.
5. The project engineer allows such friction to develop between his immediate group and other interested groups (for example, production) that present or future progress of the project is hindered.
6. The project engineer delegates parts of work to others and then fails to exercise the leadership needed to draw everything together for an excellent solution.

Chapter Fifteen

Human Relations
in an Engineering Organization

One simplifying principle, applicable to all human relations, will lay the basis for solving any problem in this area. It will show that a solution is always possible if there is any reason for the relationship to exist (although some solutions will be drastic). The principle is this: *effective relations between individuals and groups are established on the basis of a common, shared purpose.*

Chapter 3 examined the operation of a project team. Here we will be interested in taking a longer and more general look at some other aspects of the relations in an engineering organization. You will be frequently called upon to go outside your project group to deal with other groups—vendors, customers, and so on. You must deal with your own manager effectively.

We saw (even more generally) at the beginning of the book that if you are to fulfill yourself and your professional purpose you can't be a hermit or keep the rest of humanity—the people you've enlisted to serve—at arm's length. An ability to work effectively with others is essential in our business.

If the principle stated above is fundamental, we can deduce from it that ineffective relations can be caused by lack of common purpose, by the absence of a shared goal. To work with others we must thus know enough about them, their interests and motivations, to build a common purpose in whatever transactions are involved. For example, cooperative relations within a project team come from the mutual desire to make the project succeed for the benefit of all concerned.

At the other end of the relations spectrum even a man obeying another at gunpoint shares with him a temporary common interest in doing what he is asked. Their *ultimate* motivations are antagonistic. The first wants to do what he is told to save himself from harm. The second wants him to do it for some other reason. Thus the common shared purpose need not be fundamental and can spring from the highest or the lowest of motives.

But there are problems which can prevent or destroy an effective shared purpose. You must also know enough about people to understand these problems and overcome them. Human beings appear to be moved, at the same time, by many different purposes, often conflicting—for example, the desire to eat and the desire to share. Their motivation, or the relative importance of their many motivations, changes from time to time. One is not hungry after eating; he is perhaps more tolerant and generous at Christmas time. At any rate the *net purposes* of individuals with respect to some activity must be held in common for them to cooperate with each other in it.

A professional psychologist might object to a one-chapter discussion of so basic a topic and insist that it is a life study. Indeed it is, and for each of us. But you will realize that you have already devoted a good many years to this study—and successfully enough to gain an engineering degree.

Economic Motivation and Fulfillment

There are a few basic points about engineering relations which, kept firmly in mind, can make a vital difference in engineering effectiveness. Young engineers have *real* trouble with human relations needlessly. (But even the best will have problems at times.)

Look at the possibilities of motivation for a common purpose in terms of the four human needs discussed in Chapter 1. We may summarize them very crudely as (a) the economic need (the first three needs especially) which is represented quite well by money and employment, and (b) the fulfillment need, which shows most clearly in what one thinks of himself and his status. But these needs are peculiarly intertwined.

Nothing Wrong With Economic Motivation

The economic questions each of us face—"Will I lose my job if this action is unsuccessful?" "Will I be promoted if my idea works?" "Will I make a profit?" "How much will this cost me?"—these motivations are often misconstrued. They are not so selfish as first appearances would indicate. Money is the measure of all economic activity and of many others (but not all). Success or failure is usually measured by monetary standards. The superiority of a free enterprise system comes from the effect of this common measure in stimulating *almost automatically* right economic decisions in millions of places and at all levels.

What we frequently criticize as a "money-grubbing" or "What's in it for me?" attitude is really blameworthy and foolish only in that it is short-sighted. The person acting in that way is usually concerned with relatively small profits today while unconsciously shutting himself off from much more tomorrow. His actual misdeed is that he is not really taking care of himself very well at all.

If we are running our society right we should expect that right actions will bring economic reward. Thus the person who acts intelligently from an economic motive ("enlightened self-interest" was a popular term for it in past years) may not be far from effective self-fulfillment.

Appeal to Highest Common Purpose

Fulfillment requires living up to the highest and noblest ideals of which we are capable. These ideals thus appeal with a power which is surprisingly proportional to their content of practical idealism. Hence a good rule in dealing with others is to *work from the highest common purposes possible while not overlooking the economic motive*.

Dishonesty, insincerity, and the intention to manipulate others do not promote effective relations since they destroy any really high common purpose. The gunman does not command his victim's professional skill or enthusiastic, creative support.

Threats against Self-Image

Relations between individuals and groups are greatly influenced by (a) how the parties are organized, and (b) how effective "communication" is between them. But first we will consider a complete relations upset that actually occurred in engineering work. Why did it happen?

In a large company an opportunity to bid on a substantial system study and development job for an outside agency was uncovered by department X, a relatively new organization. It aggressively pursued the opportunity to a successful conclusion over severe external competition.

Although department X had been clearly operating in this general business area, it did not have experience in all details of the particular development field involved. It was confident, however, that the same technical excellence and alertness displayed in winning the contract award would carry it through to a successful contract conclusion. Apparently its customer felt the same way.

Department B, a long-established department, had previously been doing the very kind of thing required in this study. On hearing of the work, the members of this department demanded the newly won contract for themselves.

Department B was suffering from several difficulties. It had for a number of years been subjected to a rapid turnover of general managers and section heads. This had resulted in loose and somewhat vague policies which had permitted opportunities like the present one to slip by. Also, although the department was staffed with a number of technically competent engineers and specialists (some of virtually national repute), many of these people had become engrossed in technological things. They had lost sight of practical engineering application and economic contribution to their company (and to the nation). Perhaps it would be fairer to say that they were so convinced of the superiority of their company and department that they did not deem it necessary to be greatly concerned with economic matters.

Here was a situation where some constructive relationship between the two departments was needed. Department X needed the technical capability of department B. Department B needed the work that department X had won. Even if it had been able to control the contract, however, B would have required help from X, with whom the customer was by now quite enamored.

Several efforts to resolve the problem between the two departments failed. Meetings resulted in loud and heated arguments and even name calling.

Men Need the Respect and Approval of Others

One important element of the fulfillment need in a man is his hunger to know that what he is doing is important and worthwhile. He wants to feel that others recognize his contribution and respect him for his work. They must *accept* him *as he sees himself*. This is true in every kind of work and at every economic level. It is just as important to the custodian of an apartment building as to the president of a bank. Certainly every member of an engineering section and all others with whom you work are subject to this need.

The engineers and managers of department X moved into a relatively new field and contrived a proposal scheme which won over heavy outside competition. This was a fulfilling symbol of their ability and importance. They saw themselves as successfully bringing business into an expanding department. But then someone tried to take it away from them with the argument, "We know so much more than you do about this work that for the company's good we'll have to do it." Their whole fulfilling concept of themselves was *threatened*. They fought.

On the other hand, the experts in department B had never questioned their own outstanding abilities. The impertinence of a new department trying to win an award in their own technical area *threatened* this fulfilling picture they had of themselves. They could only write off the success as some kind of fluke. To submit control of this work to others would imply that department B was not really so important after all. This would raise the question of the real value of their last ten years of effort.

In the end, resolution of the disagreement had to go to a higher manager, who had little choice but to favor department X. To do otherwise would have plainly said, "Aggressive solution of our customer's problems really isn't very important after all. Just sit back with your technology and we'll bring the application in to you." Of course, he did his best as a good manager to improve morale in department B by "requiring" that certain of its people be used in the work.

The sequel was interesting. Department B supplied two project teams and several systems specialists. After a few meetings the specialists dropped out, still unwilling to work on a team effort which appeared to damage their image of themselves, although there was much they could have contributed.

One department B team leader cooperated thoroughly, produced excellent work with his group, and was clearly a mainstay of the project. But in his particular area there was no competence in department X whatever! Thus there was no image problem, no threat to this group's fulfilling concept of themselves.

The other project team's manager was still unwilling to cooperate and started to use his funding to train younger engineers. Department X promptly dropped them from the project.

This not untypical relations problem illustrates several points.

1. A major source of dispute in an engineering organization is the question, "Who will do this work?" We can see now, in terms of self-image, what that kind of problem is. We could also interpret it economically, but this would amount to the same thing. And it might somewhat obscure possible solutions.

2. Once a dispute has been allowed to develop strongly over a period of time or to break out into an emotional display, it is difficult—sometimes impossible—to settle. Many of the department B men put on the

project, after the dispute was finally resolved, were literally unable to participate. I will stress again that they were capable, experienced engineers.

3. It is always possible, in spite of the group fight going on about him, for an individual to take an impersonal and constructive outlook. Under the leadership of such an individual one of the *B* project teams turned in an excellent job. Incidentally the team leader was afterwards offered a promotion in still a third department and accepted it. After the fight is over, there is no point in suffering further from it.

4. Some disputes are inevitable when the work of an organization is so conducted that its individuals have the opportunity for effective and aggressive use of their talents. They serve a purpose in resolving otherwise answerless questions. But, of course, disputes should be relatively infrequent.

It might be argued that higher authority should have decided the product scopes of the two departments beforehand so that such a problem could not occur. Now it is true that a great deal of needless energy is wasted on these issues by poor organization. Even the best organization, however, cannot foresee and resolve every conflict of interest.

Ideally the battle is conducted on both sides to minimize harmful results and maximize useful resolvings. Practically this means that each person should argue impersonally and try to hasten the decision. When fighting there is never any point in doing less than one's best. But by being calm and impersonal during the battle you stand ready to find and accept a worthwhile compromise.

5. Another source of friction in any organization is failure of one man or one group to fill completely its function. Others will rush in to take its place. Self-deception as to what function one is actually filling can be fatal.

Line, Service, and Staff

Every person in an organization tends to live up to what he thinks his function is. If, for example, a man has the title "Manager of Reliability Engineering" he is going to expect that his unit will do the company's reliability engineering work. Anyone else who attempts to take care of this function (on a project) in another way is in effect threatening him.

The efficient project engineer who quietly sets up his own reliability man is thereby saying to this manager, "Your function isn't important or actually needed here. Your job could be reduced or eliminated. You're not actually the Manager of Reliability Engineering anymore."

On the other hand the Manager of Reliability Engineering could have shared the decision to set up such a slot within the project. He might have been asked to furnish or to help find a man for it. Perhaps he could participate in reviewing the work periodically. He would then have been better fulfilling his concept of his own function.

There is good organization and bad organization. Most companies seem to have a mixture of the two. But before you jump in to criticize, think over the alternatives, and the difficulties they will cause even if they correct some existing problems.

A common source of friction in personal relations is misunderstanding so-called line-service-staff relationships. The terms "line," "service," and "staff" are not very precisely defined, but in general a *line group* is one which is directly concerned with the product or with a major activity of an organization. For example, in an electrical manufacturing company the engineering groups which design or make motors and controllers would be considered line groups. *Service groups* (sometimes called *staff groups*) are concerned with special, necessary activities not directly related to the product, such as personnel work or plant equipment maintenance.

When you are the engineer on the firing line in a design group it is natural to suppose that service groups will recognize the importance of design work and arrange their schedules and conveniences to yours. Somehow they do not; service units are usually not project motivated. Your project group in its difficulties and enthusiasms works strenuously (possibly long extra hours) to get the work out and make the project a success. How galling to be told at four-thirty in the afternoon by the manager of a service unit (while his people are putting away their work) that you can't have approvals for some essential nonstandard components this week after all!

You ask, "Why not organize in such a way that a line manager will have everything under him that he needs to do the job?" In principle this is a good idea, but it is impractical to carry out very far. Personnel and accounting specialists, for example, cannot be economically furnished to every part of the business. And how would the line managers supervise them, without detailed knowledge and experience in these specialties?

The term "staff" is usually applied to individuals who report directly to a manager but have nobody reporting to them. The idea here is that some managers have a larger coordinating and planning job than one man can do by himself. When well-handled, an organization using this staff concept can be very effective. The manager's work gets done correctly and on time. The theory and practice of staff work have been developed to a fine point (by the U.S. Army, for example).

Poorly handled, however, the staff type of organization is usually worthless. Some large companies have been burned so badly with it that they forbid any manager to have staff assistants (a prohibition that is conveniently avoided by calling them something else). The real problem in poor staff work is that staff personnel are not given clear-cut assignments and responsibilities. As a result they make up their own in order to have any self-image at all. If a staff man has no clearly understood function, associates do not know how to deal with him and relations are strained.

Good organization, which provides not only titles and reporting responsibilities but also *specific job functions,* brings out the *technical problems* which must be solved in a company's work. It should minimize administrative interfaces and artificial problems such as the dispute between departments *B* and *X*.

Once a structure is in place the alert engineer recognizes its definitive effect on his relations. Even a man who performs a service function finds it very difficult to avoid feeling that his work is one of the most important operations in the plant. Design and production, he thinks, should accommodate themselves to it.

After all, a man's work *should* be the most important thing to him. We can hope that he also sees the overall picture of his plant's operation clearly enough to fit his work into it effectively. Robbing him of his satisfaction and self-image, however, will not encourage him to take this view.

Communication

A few years ago a number of large companies in the electrical industry were convicted and punished for violating the antitrust laws of the United States. They were accused of allowing some employees to conspire clandestinely with each other to fix prices and rig bids. Some individual managerial employees were fined and even imprisoned. Many left their companies.

In one company, for example, it appeared from news sources that this practice had been indulged in intermittently for many years, not by the company itself but by a surprisingly small group of employees attempting dishonestly to maintain their own (or their department's) sales or return budgets. Some of these people had worked into fairly high places by the time of the litigation. Some had been engineers, surprisingly enough, who must have forgotten their engineering missions!

The point of interest is this. Suppose that you are president of a company and want to be sure that no officer or employee will indulge in such activity. You know that this sort of thing (like any other piece of chicanery) must serve eventually to discredit both company and individual. How would you go about it? What would you tell your employees?

Remember that there may be a few who are sincerely, if erroneously, convinced that it would be to the benefit of the company to break these antitrust laws. They regard them as damaging to the nation.

Do you tell them firmly, "Observe the laws?" If so, the employee may say in his own mind, "He knows we have to conspire with our competition to stay in business. He wants me to do it. Of course, he can't say so and he doesn't want to know anything about it. I understand him. I get the message!"

Now you, as the president, may suspect that this sort of thinking is going on, so you double your exhortations and regularly publish a directive to all concerned to desist from any such illegal activities. What is the result? There may be those who think, "He's just making doubly sure that the company is covered in case of trouble. I know what he really wants me to do." And this sort of thing could go on indefinitely.

It is clear here that the president and his employees are conversing with each other. They clearly understand the *English words* involved, but they do not really *understand each other* at all. In fact the president is conveying, though unintentionally, exactly the opposite of what he wants to express.

This is a rather extreme example, but it illustrates that, although people may talk together, they do not necessarily get across to the other person the ideas they wish to convey. In modern terminology, the president was not *communicating* with his people.

Communicating with others, as opposed to merely talking (or writing) to them, implies that substantially equivalent *ideas are being shared* Communicating is not merely from mouth to ear; it is from mind to mind. You might almost say from heart to heart. It is easy to see that much more is involved than words. Tone, actions, sincerity, and past relations (to mention only a few items) will play a part in how words are interpreted or indeed whether they are considered at all.

In your dealings with others you will find many obstacles to effective communicating. They have to be overcome in some degree if you are to succeed. We can pretty well group them into two classes: difficulties caused by differing backgrounds between the people trying to communicate, and difficulties caused by emotion. Background problems can be as fundamental as not speaking the same language or as simple as not having been to the same earlier briefing. In between come differences of outlook. Examples are those between two men, one of whom has spent twenty-five years in production work while the other is similarly experienced in research.

Getting around Background Problems

A good procedure in communicating with anyone for any purpose is to start from some mutually understood situation or idea. Proceed from this common ground to the business at hand. Frequently it is useful, in starting the discussion, to briefly summarize where the two parties are.

For example, in contacting purchasing people to expedite an order you might say, "You remember the special cathode ray tube you helped me with on the XYZ project last month. We finally placed it on Consolidated, didn't we? And we felt they had quite a bit of unused capacity right now. Well, ..."

Or, in presenting a large project status report to a customer group, you begin, "A number of you gentlemen were present last December when we reviewed progress for that quarter. Let me summarize where we were then so we can start together in looking at this new work."

Such approaches serve to bring people together in their thinking. The following communication of ideas should not then founder on a significant difference of background. Also, if there is some important problem here, it will be brought out to be solved before proceeding further.

Emotional Interference with Communication

Emotional interference with effective communication must in a similar way be considered and removed or reduced. The principal problems here, as you might suspect, involve status and self-image.

There is a difference in the emotional attitudes that various people will take to the same situation. But observe generally that *any seeming or even possible threat* to a man's position will cause suspicion. This suspicion, with or without a real cause, can quickly grow into resentment and even hatred.

Those with genial and openly friendly dispositions are looked up to by all for, among other reasons, their very rarity. Cynics say that these people are genial and friendly only because there is no threat to their positions. A more plausible explanation is that their superior understanding of people (including themselves) frees them to take this effective attitude with others. It also allows them to establish themselves securely.

When a man is hard pressed by circumstances, whether on the job or in personal matters, he is naturally going to feel more defensive, more suspicious, more emotional. Those dealing with him will want to consider these facts and to take the extra measures necessary to communicate under more difficult circumstances.

Measures to reduce emotional interference with communicating include demonstrating that no threat exists, or showing clearly what the limits of the threat are (as, for example, in the relations between manager and subordinate if the latter is not performing well). Demonstrating personal friendliness and good will is always a help.

If no threat is involved, mutually understanding the purpose of communication should eliminate most emotional interference, but even this understanding requires some communication. Hence the real issue is approached gradually.

Another approach, particularly good where other methods cannot succeed, is analogous to raising the signal level in technological communication to get through when there is interference. In the kind of communicating we are talking about, the most effective way to "raise the volume" is to tie the message clearly to an economic result. The boss says to a subordinate, for example, "If you can't improve your relations on the factory floor over the next six months, we just won't have a place for you anymore." Or, in approaching a colleague in manufacturing with a change proposal for the product, you say, "How would you like to cut a couple of dollars per unit off your production cost?"

Where the people involved know and trust each other, there is less difficulty. Strangers communicating for the first time need more elaborate preparation. Technologically minded engineers are frequently surprised at the amount of small talk and generalities that precede getting down to business at these meetings, but we can see now that such preparation is essential. Participants during this period are "getting to know each other," that is, displaying their good will and taking the measure of others. They then proceed by this means to establish informally the fact that all are aware of some common purpose in communicating. There is no threat apparent. Then free communication actually takes place.*

Relations with Your Boss

At least the *principles* of effective relations with the boss are not much different from those involved in getting along with anyone else.

---

* But see also the last topic in this chapter.

But you report directly to him. He will write your efficiency report and control your raises and promotions. Therefore both your economic and status interests are immediately involved. Two common mistakes in working for a man are (a) to forget that he is king in his own group, and (b) to look to him for guidance in every activity.

It is helpful to think of groups directly in terms of their leaders. The Army often names temporary battle groupings after their commanders—"task force Jones," for example. In effect an engineering unit is simply an extension of the manager's own activity. What the boss wants, he should have immediately unless there is some flaw in his plans of which he is unaware. In this case his subordinates warn him of it. But if you are busy in your work, your boss will be twice as busy in his. Therefore he will usually be best supported by prompt compliance with his real wishes and needs in your area.

One common way of undermining the boss' status is to deal with outside groups, especially his superiors, without his knowledge and consent. There will be routine contacts which you continue to make without his detailed knowledge—with your drafting support, for example. It is vital, however, that anything at all unusual in out-of-group contacts be brought to his attention first.

For example, you arrange with manager *A* of another unit to support your project with additional shop work. Unless this is a routine and regular channel for you, your boss may be embarrassed when manager *A* surprises him at lunch with the remark, "Now that we're committing another man for Jones's project, you'll realize we can't get that special chassis job out as soon as you had hoped."

Of course, there will be times when you must act on your own initiative with your manager's own boss or some other outsiders. The matter can't wait, and perhaps your manager is out of town. In these circumstances your prompt and intelligent activity will redound to his credit. But you tell him as soon as possible what has been done.

Similarly keep him informed of progress and of anything unusual in your work. Don't let him be caught with a surprised look on his face when his own boss says to him, "The manager of scheduling tells me Jones's project is going to slip another four weeks. What was that vendor trouble you are running into there?"

It's his unit. The boss must know everything that is going on in it. His subordinates must take the initiative in keeping him informed, particularly on anything that pertains to outside groups.

On the other hand we have already seen in Chapter 3 the impossibility of a project engineer running every detail of his project. A unit manager is even more incapable of doing all his engineering himself. He has hired you to think and to solve his problems. If he only wanted you to do exactly what you are told, he could have hired a technician—and a fairly inexperienced one at that.

Don't run to him with every problem. When you do go to him, suggest at least one course of action of your own. If he doesn't like it he can give you some better advice, but at least he knows that you've thought about the problem yourself. He is there to help you. Don't be afraid to get his help or at least to put him in a position where he can give it if he wants to. But don't let him do your work for you.

Sometimes there are bosses who don't really know what they want but will tell you something anyway. They're busy: you come with a question; they'll give you an answer. Perhaps if they thought it out they'd say, "I'm not going to answer that for you. Go back and figure it out for yourself. If you're still stuck tomorrow I'll talk to you then." But, unthinkingly, they give you their first estimate of an answer instead.

Of course, what they really want is a good solution. If you go out and do the first thing that popped into their heads whether it's appropriate or not, you're not giving them what they want.

All of which stresses the idea that you must get to *know your boss* as well as you can. See what makes him tick. Don't be afraid of him. His bark and appearance are worse than his bite. Don't bother him needlessly, but get to know him. He'll have his strong and weak points just as you do. There is a lot you can learn from him—from his weaker points as well as his stronger ones. A lieutenant-general told me, "When I was a lieutenant at Fort– – –, the post commander put out an order to – – –. I put that down in my little book right away. I wanted to be sure not to do it when I got a post of my own!"

One time my boss was placed under a new manager who took great interest in the project I was running. He called me into his office frequently for long discussions and gave strongly worded opinions and suggestions on many parts of the work. In carrying these out the project became muddled and confused. It was several weeks before I came to realize that this was simply a mannerism on his part. He didn't intend to be taken seriously. He was simply exploring technical possibilities from his own extensive and successful experience. He expected me to use my judgment on which to pursue and how far to carry them. Get to know your boss!

In a similar way when you have other than routine business with the members of another group it is polite to approach them through their manager. This in effect acknowledges his status and puts him in a position to know what is going on in his own group. This courtesy is especially important when you are carrying out directly some instructions given by your own unit manager, so that you could get him as well as yourself into trouble by some indiscretion.

Informal Channels

There are in most organizations both formal and informal channels for doing things. For example, to get copies of prints from the print cage it may be normal to take a filled-in form to a certain office for recording and approval. But you know the girl in the cage quite well and can get copies faster directly.

Which channel should be used? Formal channels are best almost all the time. If they are clumsy or outdated, your suggestion (first to your boss) may result in an improvement.

Informal short cuts are invaluable in emergency but must be used with great caution. Normally, when informal channels are used on anything important a parallel "confirmation" should go down the formal channels so that everyone concerned is informed. Most formal channels are established for a good reason, which may not be immediately apparent.

Take variations from routine procedures when there is a really strong reason to do so, but take them with great care. At the very least you are probably threatening someone's status.

Human Relations Problems Never Change

One of the most practical and helpful sources of advice for young engineers concerning effective relations is a series of articles written for the journal *Mechanical Engineering* by W. J. King some thirty years ago, under the general title, "The Unwritten Laws of Engineering." Although engineers may have improved somewhat since that time in some of the problems that King covers, basic problems of human relations and communication will always remain the same.

The injection of engineering affairs into the usual human relations problems has a big advantage. It provides a built-in common background interest from which engineers conduct their dealings with each other. But those deeply interested in technological excellence may be inclined to ignore the need for effective human relations.

A common misapprehension is that if you get along well with everyone you have effective relationships. It is true (although there are a few exceptions) that inharmony with others means noneffective relations.

But it is perfectly possible to be on the best of terms, to have the clearest and most complete understanding with someone, and still make poor progress toward a common purpose.

For example, in a somewhat slack organization a great deal of time can be wasted in personal conversation, arriving late, leaving early, etc. Everyone may be great friends, but such relations are poor! Measured against a useful engineering purpose (unless it is simply pleasant time serving), the organization is inefficient.

Most professional people are accustomed to treat others with fairness and consideration. But man is a little lower than the angels. There is a tendency in human nature to take advantage of weakness in others. Hence you can't afford to appear weak. To be so docile and affable that others are tempted to take advantage of you is a real disservice to organizational harmony.

On the other hand we sometimes see an associate so suspicious of others that he is hardly capable of cooperative effort. Or he may be belligerent, adopting as a shield a hard-boiled, discourteous, foul-mouthed attitude toward life. If such a man is somewhat successful it is in spite of this attitude, rather than because of it.

The best approach for most of us seems to be a tough-minded, confident attitude toward our work, our problems, and our associates. Give and expect fairness and honesty and courtesy. Don't be easily upset or offended. In the rare instances when an associate appears to be acting aggressively toward you, first wait long enough to be sure. Then take whatever firm, unemotional action is required to protect yourself and the project work.

Good Engineering Practices in Human Relations

1. The engineer, while battling hard when he must, recognizes that in disputes with others it is important to maintain a calm and impersonal attitude if maximum benefits are to result.
2. The engineer recognizes the importance of a self-fulfilling image to his associates and is careful not to stir up difficulties for himself by inadvertently threatening these self-images in others.
3. The engineer recognizes how strong an influence organization patterns have on setting his relations with others.
4. Realizing that differing backgrounds can interfere with his efforts at communicating, the engineer is careful to establish a common understanding and purpose before proceeding into new business.
5. The engineer is careful to avoid informal channels in most matters

except in emergency and then covers his short cut with a parallel action through regular channels.

Poor Engineering Practices in Human Relations

1. The engineer feels that his expert technological knowledge will eventually solve all relations problems.
2. In disputes with others the engineer uses kid gloves to improve relations and group harmony.
3. Not understanding the concept of communicating, the engineer naively assumes that since his colleagues understand the English language they will comprehend what he is trying to tell them.
4. Heedless of the effect of emotions on communicating, the engineer leaps headlong into a delicate situation with a minimum of results.
5. The engineer takes his boss for granted.
6. The engineer assumes that if his relations with others are harmonious they are necessarily effective.

Chapter Sixteen

Engineers
and the Marketing Function

Whatever you do, you'll need to know a lot about marketing. In a small organization it is difficult for any member, engineer or other, to overlook the significance of marketing activities or to badly misunderstand their nature. But as companies grow larger, more and more employees lose sight of marketing and have little or no customer contact. Attitudes become warped; work suffers.

A small but glaring example of this came to light in a large company a few years ago. This manufacturer employs more than two hundred thousand people, scattered across the United States. Thousands of different products are made at more than a hundred locations. A customer in a southern city, desiring to order a certain replacement part, not unnaturally called the first place he could think of, the company's plant in his city. Of course that particular plant made some completely unrelated line of equipment.

The unfortunate young lady who took the call informed the customer, apparently in no uncertain terms, that her plant did not make what he wanted and that this was not the place to order it. Then she hung up. It was apparent that she did not think in terms of her company's purpose—to serve its customers. Her viewpoint was limited to her own plant's production problems, or perhaps even more narrowly to her own job.*

It is impossible for the man who is really practicing engineering to misunderstand his company's marketing effort or to be out of sympathy with it. We saw at the very beginning of this book that an engineer's mission is to better meet human need by the intelligent application

---

* Needless to say, internal company publications made the most of this story as a bad example. The result was a wonderful series of opposite examples of people going far out of their way to help customers, particularly those who wanted to buy company products.

of technology. A company can hardly be intelligently apprised of human need or be working on it without some systematic contact.

Sometimes our newer engineers, those who are just getting started in their practice, are dubious of marketing as a way to make a living. Perhaps it is even true that usually—but not always—the more talented a man is technologically the less sympathetic he may be to marketing. But sales engineering is a highly creative vocation for many good engineers.

Whether or not a man trained as an engineer is interested in devoting a substantial part of his time to formal sales work, an attitude of indifference to his company's marketing tends to isolate him from his overall engineering function. He becomes a "responsive" engineer.

What Is Marketing?

Perhaps a major source of difficulty in understanding the marketing function of a business is the long exposure to huckstering that we all have while growing up. Magazines and newspapers are filled with a wide variety of advertising, some of which may be of benefit to the public, but more of which seems to have little or no relation to the general good.

It has been my experience that youngsters who are potential engineers—with their literal-mindedness and interest in specific facts—are more offended with consumer advertising than most people. Not infrequently they build up a prejudice against all advertising and even the marketing functions in general. Like most prejudices, this one is at least partially unreasonable, as we will see.

To carry this point a little further we are even less impressed with radio and television commercials than with printed advertising. Some are more or less deceptive efforts to induce us to believe what someone else wants us to think. Persons describe products that they probably know nothing about and have never even seen with a beguiling charm and "sincerity." It would appear that, the less value a product has or the less claim to distinction over others performing the same function, the more blatant the din raised over it by its advertisers. Most of the public soon learns to discount these "messages" and to suspect any "information" related to commerce.

But should we really discount all advertising because of the excesses of some? If you don't think that advertisements are worthwhile, why do you read them? We all instinctively identify an occasional advertisement as good, particularly if it combines the elements of (a) a lack of the aggressive content objectionable in many consumer advertisements, (b) some simple and logical combination of facts, and (c) a product that we know or feel is indeed excellent. We are willing to see others sold and to be sold ourselves on a product that in our opinion will truly fill a need, if that selling seems to us honest. We are glad to have the information that such a device exists and to know how it can be applied for our own use.

Our experience with useful advertisements, though some may think it quite infrequent, serves to indicate that there is a function to be performed here. We need to bring the business (in our case an engineering-oriented business) into contact with the needs (into contact with people whose needs we wish to serve). And that, of course, is the *function of marketing: to uncover the needs that we engineers will better meet, to bring about a fruitful contact between the user and the supplier of this need-filling activity.* Unless this function is provided for in some way there can be no engineering.

There are different ways whereby the marketing function can be handled. Advertising is just one and is used in conjunction with others. The engineer may go out and discover the individual need for himself. He figures out how to provide for the need, shows the prospective user the advantages of his new idea, and arranges to supply his need-filling product.

In a second kind of marketing approach the engineer investigates a general area of need carefully—say the need for a certain kind of agricultural machinery. He designs a line of such machinery for use by many growers. He then goes out and acquaints them with his products and their advantages so that the growers can buy and use them (thus completing the engineering cycle).

Or the engineer may associate himself as a consultant with the users of his art. He sells himself on the basis of his previous experience and reputation.

We saw in Chapter 15 a typical breakdown for an engineering-oriented business in which the formal marketing function was organized separately with its own manager. This is an efficient division of labor. Some people put their major and specialized efforts on uncovering needs and providing for the mutually profitable association of users and suppliers. Others specialize in design or development or production. We will look at this kind of marketing organization in more detail shortly.

Finding the User or Customer

Chapters 8 and 10 discussed what appears to be the chance nature of much design and development work. Because of this element of chance it is impossible to forecast how this work will come out in detail. Neither developer nor designer can predict how his design details will turn out. Each makes it a point to search for useful new ideas. Each puts himself where he is likely to run into these ideas. Recognizing the seemingly chance nature of his business, he strives to improve his odds.

In a not dissimilar way your marketing organization is faced with a probabilistic situation when it looks for customers in a rapidly changing technological society. This is true if the business is to uncover individual needs and to tailor-make solutions for them. It is just as true if the business undertakes to fill some more general need with a line of products. Marketing personnel, especially the salesmen, recognizing the chance element in their work, improve their odds by gaining as wide an exposure as possible in likely customer areas.

They are diligent and persistent in searching out people and organizations that the company can serve. Their zeal is similar to the development engineer's diligence and persistence in seeking a feasible and economic solution to his part of the engineering problem.

In engineering—meeting need better by the intelligent application of technology—the idea that success is *service* is a particularly clear one. Thus the sales volume of a company's products is a good measure of its success. If marketing people are to serve effectively, they must have a thorough knowledge of (a) their products, and (b) the industry to which they are selling.*

Subdivisions of Marketing

The marketing function of a modern, engineering-oriented business may include some of the following:

> Sales.
> Product planning.
> Market research.
> Advertising and sales promotion.
> Marketing administration.
> Product service.
> Contract administration.

Let's look briefly at each of these.

---

* According to a marketing manager of a medium-sized valve company, every one of his salesmen should honestly convey to the customer this impression: "I've taken the time to learn my product and how it applies to your industry."

Sales

The sales group is, of course, the heart of marketing effort. From a position right on the firing line they bring in customer orders. All marketing effort is organized to assist and support sales. The effectiveness of a sales unit's work is conveniently measured over a reasonable period of time in terms of dollar-volume of orders received. Over a relatively short period, however, this measure may be quite misleading.

During a season when the company has some substantial technological advantage over its competitors it is easier to get a large volume of orders. But when the product is in technical or production difficulties, salesmen are at a real disadvantage. Therefore sales effectiveness in terms of orders must be measured thoughtfully over a period that is long enough to minimize such effects.

To the engineer, sales organizations seem to be in continual change. Their work is organized to cover (a) the customer areas required (geographically or by type of industry served), and (b) the range of company products. Changes in these two reasonably require changes in sales units. In addition many sales managers seem unable to decide which kind of organization to emphasize. Small companies with a limited product line put their emphasis on geographical division. A convenient combination of the two, particularly for larger companies, includes district sales offices near concentrations of customers plus a product division of work for the home office sales force. In theory, district office personnel first uncover possibilities for company service, and more specialized home-office personnel then assist in completing the sale.

There are many variations, intentional and unintentional, to these patterns, but we will not be interested in details here. Recognize again, however, that without this effective customer contact there can be no engineering. To serve people through the better application of technology to their problems, you must know about these problems. You must know who the people are and where they are, and must bring your technology or your product to bear on their situation.

Your company may have an outstanding technical product, far superior to anything else available. But you are not going to be of much service to the customer (and to the economy and the nation) if few potential customers know about your product. Also, if most of those who would be interested do not understand how it works or how to apply it, you'll be in the same trouble.

Product Planning

The other subdivisions of marketing are usually quite subordinate to sales but are not without interest and importance to the engineer. Product planning in theory is done by a group that carefully looks at both market needs and company capabilities, to come up with new product ideas. In most of the organizations that I have been associated with or have observed closely, this function is hardly anything more than a pious hope. It is most ineffectively handled.

New products are the lifeblood of almost every business and hence a primary concern of top management. If product planning is to be effective it cannot be relegated to a couple of marketing men and forgotten by everyone else. Where the function *is* effectively handled the so-called product planners act more as a clearing house and stimulant for new ideas *from all parts of the organization.* The sales group in particular (with its daily customer contact activities) should be a fertile source of new ideas.

Research and development engineers, or the "advance engineering" group as they are often designated, must be closely associated with the product-planning function. In some organizations with a heavily technical product line development engineering works under the manager of marketing instead of the manager of engineering. Although this can have disadvantages, it does bring new engineering knowledge directly to the product-planning function and guides the work of the developers themselves into more immediately useful channels.

Market Research

Market researchers dig out and assemble statistical facts, which sales and product planning use in their part of the business. How many industrial plants are there that could be customers for the new product? Where are they located? What are the sales figures and organization of competitors in similar lines? What has been the trend over the past ten years in the use of our company's products? How does it extrapolate? Answers to these kinds of questions can assist budget planners in all parts of the business. The manager of sales uses them in assigning his people and setting up district offices. They help to guide product planners and development engineers.

Product Service

Where a company's products are complex technically, a product service organization may be needed to assist the customer in various ways—training his people who will use the product, trouble shooting, maintaining and repairing. A radar supplier to the Air Force, for example, may have product service people (techreps) at Air Force bases around the world. The product service group may establish repair depots for company products. (In some businesses this function may be assigned to the production group, although it would seem to be more nearly related to marketing than to production.)

Contract Administration

For some businesses orders result in involved and extended contractual relations between customer and supplying company. (Engineering development contracts and government orders are good examples.) It is sometimes efficient to relieve the sales group of responsibility for continued customer relations once an order has been brought in. In this case a contract administration unit takes over the order as soon as it is booked, handling the paper work and administrative details. The members of this unit act as customer liaison for the whole project. They may set up a monitoring and reporting system with conception-design engineering and production to be sure that deadlines and report requirements are observed. Any special accounting procedures required by the customer are monitored.

Engineers in either design or production may then find themselves working closely with contract administration on a project. In one sense these administrators become an engineer's local customer—often a helpful situation. But the engineer must make sure that he is also considering the needs of his real customer, and not allowing the contract administration support to filter out the customer contact that he requires.

Engineers and Customer Contact

Working with the customer on a going or a potential project can be a most stimulating kind of engineering. Sooner or later almost every engineer will be given the opportunity to do this.

Salesmen or sales engineers, with or without a technical background, frequently need strong technical help in dealing with the customer's problems. Also, they must convince the customer of the technical adequacy of company products to meet his need. Often individuals who will use the product in the customer's organization and who will make or influence the decision to buy it are engineers themselves. They sometimes prefer to deal with other engineers in the matter.

Sometimes capable engineers are taken along on customer visits simply as ornaments. Even this situation can be interesting and worthwhile. If the customer is going to let a company apply its technological capabilities and products to solve his problems, he must first have some reasonable faith. By intelligently discussing these problems and possible solution approaches the engineer can help establish this faith. In the meantime he himself is also learning about typical problems to the resolution of which his company can contribute.

Don't Torpedo Your Salesman

There is a technical side to marketing that many engineers are hardly aware of. In place of the engineer's technology it is concerned with contracts, terms, and a knowledge of commercial practices in the industry involved. It includes an understanding of the present competitive situation, a perspective on the relation of the current order possibility to past and future business, and many other factors that are often quite subtle. While the engineer works directly with the customer's engineers on technological problems, the salesman often works out his kind of problems with the customer's buyer or purchasing department.

The engineer will not expect the salesman to interfere in the conception or design work of an engineering unit, although there is a close interplay between them on ideas. In the same way he will want to be careful (in assisting with customer work) that he does not interfere with marketing plans and strategy.

Prices and terms for most nonconsumer products are established by negotiation between the buyer and seller—clearly a marketing function. This process is a major element in making the economy of the country run smoothly and efficiently. When he is included on a proposal presentation team or other customer work, the effective engineer informs himself of plans and strategy and avoids weakening his negotiators' position.

The design engineer and the marketing man can often help with suggestions and effort in each other's area of responsibility. But there can be only one person calling signals for a team effort in critical situations.

Handling a Development Customer

Once a project for a particular customer has begun, customer relations work can fall more and more on the engineer. Inexperienced engineers sometimes make the mistake of trying to hold customers at arm's length. They give them as little information as possible. In most cases it is helpful to do exactly the opposite.

Take the customer on board; recognize that it's his project too. Your successes are his successes. Your troubles are his troubles. Give him all the credit for assistance that you honestly can. You are teamed with him to do a job. Put your best foot forward. Avoid burdening him with details, but don't try to hide major problems from him. With this kind of treatment he should develop confidence and be willing to leave the work to the people who are doing it.

Infrequently a customer's representative will become a nuisance, demanding more and more of your time and attention, and sticking his nose into every detail of your work. This has to be stopped in one way or another. Fortunately in most cases this intrusiveness is only a symptom of some trouble *he* has. Find out what the matter is; put yourself in his place.

If his superiors require an extensive periodic report on your project, help him to obtain the information and see that he gets it straight. If he is bored, arrange some noninterfering work for him. If he is concerned about the impression he's making on his superior, maybe you can include his boss in the distribution of some nonconfidential reports. But whether these measures help or not, he can't be allowed to disrupt the project effort.

Our marketing friends tell us that the trouble with most engineers is that they don't understand customers as people. They should realize that customers have troubles, too. For example, customer deadlines that the marketing people bring into the plant are often vital. Maybe without this delivery the customer's operation could be shut down with consequent expense and probable loss of future orders.

The cliché "The customer is always right" is both true and false. It is false in the sense that he should be allowed to dictate technical solutions and mar engineering work. But it is entirely true in the sense that the purpose of engineering work is solely to meet some real need. Most of us err on the side of not paying enough attention to our customers rather than catering to them too much.

Don't Fall into This Trap

In his relations with his own marketing organization the engineer may encounter a problem similar to this one. He participates in proposal work to do a project for a certain customer. His company is awarded the job, and he begins the work. As the project develops, the customer begins to demand more and more. Even where there is a tight contractual specification, changes are requested or hinted at or talked about.

The engineer can see the overall desirability of these changes but recognizes that they demand more technologically than the original plan. Although he thinks that they might be feasible, he is naturally reluctant to jeopardize the project in trying for them. But his marketing associates (in their understandable desire to accommodate a customer) press the engineer to change his approach. Perhaps they even see a glimpse of some far greater application potential if the changed project is a success.

If the engineer refuses to agree to changes which will add substantial risk to the company's work, he stands alone against the customer and his own sales associates. If he yields to their pressures and enthusiasms and solves the new technical problems, everything is fine. But if the project team is unable under the changed situation to find a practical solution to its problems, it is the engineer and his group who have failed. The fact that changes were requested is quickly forgotten. More importantly, the engineer feels that his agreement prevented his company from making the contribution that might have been possible with the original plan.

Two procedures are needed to prevent this kind of problem or at least to minimize its effects on the company. First, all contractual relations must be defined as specifically as possible. Where certain items cannot be specified, this fact should be noted with some appropriate provision for determining how they will be handled.

There are today many engineering projects in an early development stage (especially for defense and space) where the only reasonable goal is to do as much as possible. Some form of cost-plus-fixed-fee contract is usually written to permit this kind of goal. Even here the problem of changes is encountered. In place of specific contractual requirements the customer frequently substitutes his *expectations*. These can easily be escalated by careless marketing people. The customer's judgment on the success of the project and his interest in continuing it can be seriously warped.

A second procedure will prevent this last trouble. The engineer has the best grasp of technical possibilities. He must be sure that he makes the technical decisions on his own job on the basis of his own technical judgment. This does not mean that he will refuse to consider possible changes and improvements, but he can't afford to let anyone else lead him around by the nose in these matters. When you are in this situation make sure that your marketing counterpart understands your plans and signals and does not yield to temptation. Don't let the customer raise either his contractual requirements *or his expectations* without your consent.

Yielding to customer pressure has another undesirable feature. The next time that customer will be even harder to deal with. All members of your company should be realistic with their customers all the time.

Challenge in Technical Marketing

Some of the engineering possibilities in customer problem-solving work—in marketing—are brought out in the following experience.

A particular customer had spent considerable money to develop a certain piece of airborne electronic gear. The development had gone through many difficulties and had recently been placed in flight test at an airframe manufacturer's test field. All flight tests failed completely, and the customer set up a large meeting to announce cancellation of further development work on the project. The engineering company wanted to go on with the development if it was really feasible but was willing to drop it if no better success could be had.

A new group of engineers was put on the work. These men first spent several days at the test field and talked to everyone who had had any part in the test work and evaluation. They discovered that the equipment had always failed before test altitude was reached so that there had never been any operational test. Further probing of the equipment showed that with one exception the difficulty had always been the same—a failure in solder joints. There were thousands of joints in each complex equipment. It was apparent that the aircraft installation permitted the equipment to be subjected to extremes of temperature. It seemed reasonable to guess that temperature shock might be causing this joint trouble.

The engineers next went to their own production facility and examined the process by which the joints were made. They analyzed with production and design experts some joints which had failed and been returned. They discussed various possibilities for improvement. A modification procedure was agreed upon.

Armed with this information and plan, the engineers attended the customer's cancellation meeting. They convinced customer representatives that no real test had ever taken place other than a test of the solder joints under environmental stress. They freely acknowledged the difficulty but presented their comprehensive plan for correcting the joint design. The customer granted a three-week extension to put one new equipment in flight test. The flight test was reasonably successful, and development work continued. Later the company was able to ship a large production order of these equipments.

In engineering sales work, selling is basically a matter of finding the customer's problem and helping him to solve it. It isn't trying to get him to buy something that the sales engineer wants to sell and the customer resists buying. A frequent problem in buying is to find anything at all that is reasonably adequate. The marketing man who can bring a customer something which will solve his problem can be doing the finest kind of creative engineering.

Good Engineering Practices in Relation to Marketing

1. The engineer recognizes that his work requires interest in and response to marketing considerations.
2. The engineer, in working with a customer, takes the attitude that they are both concerned with the same problem and dealing on a basis of mutual advantage.
3. In customer contact work with sales people, the engineer recognizes the mutual advantage of establishing a good image of the company and its capabilities.
4. The engineer avoids undercutting his salesman's negotiating position with the customer.
5. After work has begun, the project engineer is careful to avoid allowing a customer to escalate either his contractual requirements or his expectations.

Poor Engineering Practices in Relation to Marketing

1. The engineer feels that his job is to develop or design or produce what he is told to, and that customer or market factors are simply an unfortunate detail.
2. In his customer work, the engineer feels that his purpose is to deceive, entice, or coerce the customer into accepting his company's product.
3. The engineer feels no need to penetrate the mysteries of marketing organization.
4. The engineer sees no sense in product-planning effort since things in the future are never certain anyway.
5. The project engineer regards contract administrators as nuisances to be gotten rid of as quickly as possible.

Chapter Seventeen

Professionalism,
Self-Development, Education

Your engineering career is like an engineering project in many ways. Perhaps the goals of this personal project are a bit more general than we would like to have for most of our projects, but they are none the less clear. You can derive them from the definition in Chapter 1 of engineering itself. Each engineer in his career is seeking to maximize his contribution toward better meeting human need through the application of technology. It is evident that there are three elements in this maximization.

First, the *technology* being applied must be more completely and usefully understood. As the engineer develops areas of expert specialization, he will want to delve deeper and deeper into their technology. He finds out more about the technology of immediately adjoining areas with which he will frequently interface. He keeps up with general advances and changes in his profession.

During his career he will move several times from one specialty to another and find all over again the need, as an efficient student, to acquire deep competence in his new technological interests. Also, as we saw in Chapter 12, most engineers will broaden their interests to include systems understanding and at least some systems competence.

Second, engineers need to better develop their understanding and appreciation of *human need.* With increasing technical knowledge they are in a continually improving position to see creatively what needs to be done and what can be done.

But improved technological knowledge without corresponding depth in really understanding need will force the engineer to wait for others to suggest applications (although the nontechnical man is in a poor position to conceive them). In your profession you are thus required to be an ever more perceptive and sympathetic member of society.

Third, engineers are continually striving to improve their techniques and performance in putting the technical answer and the human need *together*. This includes again a deeper understanding of men. Specifically it calls for an understanding of business organizations, effective management practices on the part of the engineer himself, and many other kinds of human activity beyond technical engineering.

Your own effectiveness in carrying out technological solutions for human need depends in the long run as much on what kind of person you are as on what you know. Therefore maximizing your effectiveness as an engineer will include self-development in the broadest and highest sense.

Vannevar Bush, a renowned engineer and one time advisor to the President said, "I would say to the young engineer, roam widely and dig deeply on both things and men, that you may rise to true eminence in your profession. But I would also say that there are matters of the spirit, aspects of the aesthetic appreciation of a complex environment, which you should not neglect if you would lead a life of full and genuine satisfaction and accomplishment."*

Your career includes a number of contributors, as does almost any modern engineering project. Superiors, associates, customers, clients, educators—all these and others work together under your own systematic and expert coordination. As project engineer, you want to accomplish career goals. Unlike the situation on most projects, however, here all these contributors are part time. But we have seen that it is almost always essential that the project engineer be full time. Unless he can devote a very substantial amount of his time to thinking about and planning for the completion of his project as a whole, its detailed parts will not add up to a satisfactory solution.

With all his individual contributors working part time—and a very small fraction of their total time at that—there is no question at all that this engineer will be substantially on his own in integrating his project—his career. The contributions toward integration that we have requested from individual contributors in Chapter 3 can (in this exceptional case) be only limited. Perhaps his wife is his one nearly full-time contributor.

Would the engineer who leaves his own career to chance do the same for any other project for which he had responsibility? Or would he assume that someone else, a series of employers for example, could run his project for him on a part-time basis?

What would happen to the ordinary engineering project if neither the project engineer nor anyone else on the team thought out limiting

* From a centennial address at Worcester Polytechnic Institute, October 9, 1964.

factors or decided how to convert them to strategic factors? What would happen to a project if no effort were made to include the scheduling, monitoring, and controlling functions? If a project is long range and a bit nebulous, does this call for an abandonment of control efforts or for an even more careful attempt to compensate for these difficulties?

Professionalism

The term "professional" in its broadest meaning includes anyone who follows some occupation for livelihood or gain. More and more often, however, it is being given the higher connotation of someone who practices a *learned* profession (a) essentially *under his own direction* and (b) *in an ethical manner* for the *benefit of others*. Let us use the word in that sense.

It is useful to contrast the concept of professional with one who works primarily for gain and at the immediate direction of others. Presumably this nonprofessional has little concern with exactly what he is doing and how he is doing it so long as it pleases his superior and thereby earns him his immediate wage. (However, all activity can partake advantageously of some qualities of professionalism.)

Creativity· is an essential element of professionalism.* Mediocre, repetitive work neither meets the requirement that a professional work under his own direction nor the requirement that an engineer *better* meet human need. It is important to recognize that the concept (or better, the attitude) of creativity applies to all demands of engineering work and is not limited to the technical.

Engineering can be practiced effectively only as a profession. It obviously requires service to others—that is its very purpose. It rests on deep learning in physical technology which includes laws and principles that cannot be violated. For this reason, as well as the service concept, it must be practiced ethically. One can hardly be honest with nature and dishonest with people at the same time.

You've got to be strong enough to discipline yourself to undergo the formidable technological training needed. The practice of engineering also demands the courage to accept essential ethical standards. In addition, you must be able to proceed on your own. We saw, for example, in Chapter 3, that the complexity of engineering makes it impossible for one man to direct efficiently in complete detail the efforts of others.

William E. Wickenden, in his book *A Professional Guide for Young*

---

* Chapter 18 discusses this topic more fully.

Engineers, lists six characteristics of a profession. With a little thought these might be derived from our understanding of what engineering is.

1. Renders a specialized service based upon advanced specialized knowledge and skill, and dealing with its problems primarily on an intellectual plane rather than on a physical or a manual-labor plane.
2. Involves a confidential relationship between a practitioner and a client or employer.
3. Is charged with a substantial degree of public obligation by virtue of its possession of specialized knowledge.
4. Enjoys a common heritage of knowledge, skill, and status to the cumulative store of which professional men are bound to contribute through their individual and collective efforts.
5. Performs its services to a substantial degree in the general public interest, receiving its compensation through limited fees rather than through direct profit from the improvement in goods, services, or knowledge, which it accomplishes.
6. Is bound by a distinctive ethical code in its relationships with clients, colleagues, and the public.*

The degree to which any specific engineering practice is good or bad professionally can be quite well determined by considering the above list. Should an engineer in his practice act as consultant on the structural design of a certain bridge? If he has the specialized knowledge and skill required, yes. If his special knowledge and competence do not include this type of bridge, no. Does an engineer attempt to sell his own idea by casting discredit on a colleague's capabilities or reputation? Certainly not. If in the course of his professional duties an engineer should render an unfavorable technical judgment on the work of another engineer, will he do it? Yes, this is certainly part of his duty to his client and possibly ultimately to the public.

Helpful discussions of many engineering ethical problems are available. Although each practitioner must decide for himself the ethical issues of his own work (unless he is so remiss as to incur the disfavor of state registration boards or national engineering societies), there is no lack of helpful literature to guide him.

One of the best ethical summaries available is "Code of Ethics of Engineers," formulated by the Engineers' Council for Professional Development (ECPD)*.

---

* Used by permission of The Engineers' Council for Professional Development.

The ECPD has organized their ethical recommendations for engineers into three parts: The Fundamental Principles, The Fundamental Canons, and Suggested Guidelines for Use with the Fundamental Canons of Ethics. The first two parts are reproduced in their entirety on page 212. The third, more detailed part can be procured from ECPD and many technical societies. In the paragraphs that follow enough of these "Guidelines" have been reproduced to give the general flavor. Perhaps those selected here will be of particular interest to men and women just starting in the profession.

✿   ✿   ✿

Engineers should seek opportunities to be of constructive service in civic affairs and work for the advancement of the safety, health and well-being of their communities.

Engineers shall be completely objective and truthful in all professional reports, statements, or testimony.

Engineers shall not maliciously or falsely, directly or indirectly, injure the professional reputation, prospects, practice or employment of another engineer, nor shall they indiscriminately criticize another's work.

Engineers shall not pay nor offer to pay, either directly or indirectly, any commission, political contribution, or a gift, or other consideration in order to secure work, exclusive of securing salaried positions through employment agencies.

Engineers shall not solicit nor accept gratuities, directly or indirectly, from contractors, their agents, or other parties dealing with their clients or employers in connection with work for which they are responsible.

When, as a result of their studies, Engineers believe a project will not be successful, they shall so advise their employer or client.

Engineers shall continue their professional development throughout their careers, and shall provide opportunities for the professional development of those engineers under their supervision.

Engineers shall endeavor to extend the public knowledge of engineering, and shall not participate in the dissemination of untrue, unfair or exaggerated statements regarding engineering.

Code of Ethics of Engineers

THE FUNDAMENTAL PRINCIPLES

Engineers uphold and advance the integrity, honor and dignity of the engineering profession by:

I. using their knowledge and skill for the enhancement of human welfare;

II. being honest and impartial, and serving with fidelity the public, their employers and clients;

III. striving to increase the competence and prestige of the engineering profession; and

IV. supporting the professional and technical societies of their disciplines.

THE FUNDAMENTAL CANONS

1. Engineers shall hold paramount the safety, health and welfare of the public in the performance of their professional duties.

2. Engineers shall perform services only in the areas of their competence.

3. Engineers shall issue public statements only in an objective and truthful manner.

4. Engineers shall act in professional matters for each employer or client as faithful agents or trustees, and shall avoid conflicts of interest.

5. Engineers shall build their professional reputation on the merit of their services and shall not compete unfairly with others.

6. Engineers shall associate only with reputable persons or organizations.

7. Engineers shall continue their professional development throughout their careers and shall provide opportunities for the professional development of those engineers under their supervision.

Engineers shall issue no statements, criticisms, nor arguments on engineering matters which are inspired or paid for by an interested party, or parties, unless they have prefaced their comments by explicitly identifying themselves, by disclosing the identities of the party or parties on whose behalf they are speaking, and by revealing the existence of any pecuniary interest they may have in the instant matters.

Engineers shall not knowingly undertake any assignments which would knowingly create a potential conflict of interest between themselves and their clients or their employers.

Engineers shall not accept compensation, financial or otherwise, from more than one party for services on the same project, nor for services pertaining to the same project, unless the circumstances are fully disclosed to, and agreed to, by all interested parties.

❀   ❀   ❀

One of the most obvious differences between professional and nonprofessional attitudes involves the relationship with employers. The professional makes himself an enthusiastic member of whatever cause or organization he is supporting. If he is employed by company *A*, he wholeheartedly supports company *A*. If *B* is his client, he can be trusted to handle *B's* affairs with the same discretion and expert attention as he would his own.

On the other hand, if he cannot condone company *A's* policies or actions, he will disassociate himself from them at once rather than continue drawing a salary for half-hearted support. If *B's* actual or expressed interests conflict with those of another client or cause that the engineer is supporting, he will not accept *B's* commission. Contrast this attitude with the nonprofessional one of accepting whatever task will pay a desired remuneration, whether one can identify himself with it or not.

Personal Development

We examined briefly in Chapter 2 some of the strengths and weaknesses of the engineer as he emerges from college into practice. Although his pattern of pluses and minuses is different from that of other professions, the general beginning is the same. A physician starts his work with an internship. The lawyer enters the office of a law firm for similar apprenticeship. An army officer enrolls in an artillery or other specialized school. The engineer also begins somewhere (informally) as an apprentice to those more experienced in the profession.

Thus an engineer faces at the outset of his career a period of intense personal development. We noted in Chapter 2 that a good rule is to make two efforts in personal development. Aim one at developing, maintaining, and expanding a special competence in one or possibly two areas. Aim the other at keeping thoroughly abreast of the profession as a whole and its major developments.

It is important to note that both these efforts, but particularly the second one, require a growing personal breadth as well as technical competence. It is this personal development or human education which enables a man to *apply* his technical competence effectively. Personal breadth allows him to see his overall problem and the many day-to-day detailed problems in proper perspective. Personal growth enables the engineer to understand others so that he can work effectively with them, serve them, and later lead them.

Although this intense self-development at the start of an engineering career may appear to ease off after a few years, it can never really stop if you are to continue excellent engineering. Both technology and human need will change and grow with time, and your responsibility and breadth of effort will expand with your career.

Specialization and Enthusiasm

We will look first at technological development. An engineering school graduate has a good background in technology, possibly the strongest he will ever have during his career, but he usually knows comparatively little about the areas in which he must specialize for his new employer or associates. Selecting an area of specialization is a combination of his own desires and the particular jobs that are open to him.

Your own interests guided you originally in associating with a new company. For most engineers the *exact* specialty is not a matter of great importance. You will do as well in one as in another if each lies in one of your broader areas of interest. Beginning engineers do not know enough about engineering specialties anyway to judge very well between them without further experience.

But as an engineer gains experience he will want to be sure that his work is really interesting to him. It is often said that no one does anything really well that he doesn't like, nor can he really like anything which he does not do well. If you find yourself after a reasonable trial in work which does not arouse real enthusiasm from you, try to think out the reasons why you react as you do to the job. This may take considerable effort and discussion with others. The problem may not be the job so much as some mistaken practice or attitude in which you are indulging. If that is the case, simply moving from one kind of work to another may have the same disappointing result.

Always, however if you are not enthusiastic about your work, something needs to be done. Take practical steps to diagnose your trouble and to correct it, either on the present job or a new one. Most managers are, understandably, just as interested as you are in this problem of lack of enthusiasm, and good ones can be an invaluable help in solving it.

Technical Judgment

Although the new engineer continues to add to his store of book learning on technological subjects, his first big increment of technical self-development comes in the area of "judgment." Technical judgment is hard to define precisely but means generally *understanding the practical limitations of mathematical and physical models in engineering work.* It is largely a matter of intelligent experience.

The man with judgment knows pretty well what elements of a situation can be neglected in analysis and which cannot. To develop judgment an engineer seeks rapid and intense experience. In addition to the things that he does himself, he tries to learn from everything that goes on around him. What he hears from experienced people can help a little, but there seems to be no way to develop judgment effectively through formal study.

All your life you will be improving your judgment of technical and nontechnical situations. You will discover that for each new job or situation you must be particularly diligent in the beginning to find its peculiarities and understand its dimensions.

Deepening and enlarging your grasp of technology can also be considered an extension of your formal education as an engineer. We will consider this topic further in a later section.

Human Education

For most engineers the need for large doses of human education is such a desperate one—we saw why in Chapter 2—that it can hardly be overemphasized.

A very successful engineer who formed his own manufacturing company several years ago and is now grossing more than ten million dollars a year said recently, "When my engineers come for advice on what they should do to develop themselves I tell them, 'Go down to —— (a local arts college) and get a B.A. Or go up to the Museum and take some art classes. Or earn a degree in economics.' I tell them to stay away from the technical societies, where they'll just meet more engineers. They think too narrowly. They don't have enough imagination. They need to think about something besides engineering."

It is difficult to go along with this experienced man on the value of technical societies, but certainly there are some young engineers who would be better off at symphony concerts if they had to choose. Although technical societies would seem to be essential for further professional growth, full personal development is even more important.

In addition to further formal efforts you can obtain a great deal of human education by experience if that experience is fully savored. Engineers who refuse to participate in any phase of their projects or company activities except technical problems are closing doors on their own future.

Some companies, recognizing this need, actually *require* their professional employees to participate in some community activity outside their plants—political work, Boy Scouts, service clubs, community improvements. Almost all companies *encourage* such participation.

This policy causes some resentment among employees at times. But it seems an effective way to help people along lines where they will develop the least on the job. At the same time it can build community good will for the company.

In many career paths, including some in engineering, experience at lower-level jobs is not adequate preparation for the next higher job. Most managers need more breadth of outlook than they will develop by filling subordinate positions. Who will be promoted? Who is ready to take the next step in increasing his opportunity to fill human need, to engineer? The man who has had some human development experience elsewhere.

Self-Appraisals

In your important work as project engineer of your own career the monitoring* aspect of project control is critical. Regular self-appraisal will include progress in (a) *personal development*, (b) broadening and deepening your *technological understanding*, and (c) the *practice of engineering* to include your significant contributions to meeting human need (and the prospects that your present activity will increase them). New Year's Day is a traditional time for this regular monitoring effort.

---

* See Chapter 4.

If you are fortunate enough to have a manager who gives you regular, conscientious appraisals, you may make your own around the same time. But make yours before the manager makes his, rather than afterwards. Then you can compare his evaluation with your own independently.*

Appropriate controlling action resulting from these appraisals may be as simple as modifying your reading list and scheduling a new course in an area shown to need improvement. Alternatively, it may be as drastic as asking for a new assignment or even looking for a new position. A good manager is a wonderful help in planning improvements. But the engineer will sometimes need the courage to leave a comfortable, mediocre position when his self-appraisal shows that he is contributing below his capability.

Reading and Study

In addition to efforts to gain the right kind of experience and get the most out of it (for both technological and personal development) you need some more formal means.

Systematic, habitual reading is the single most useful method for formal learning and usually a very pleasant one. Reading can range from detailed careful study of a technical or cultural subject to rapid perusal or even scanning to get major ideas; methods are suited to material and purpose. Technical journals offer some of the best material. Every engineer usually has several which he reads regularly and a larger number which he sees less frequently. By following one or two general publications in your field of engineering regularly you are unlikely to miss completely any significant development.

For reviewing old material or learning new, books are more useful than journals if there has been time for them to be written and published. Books are especially useful for nontechnical reading, on topics where material changes more slowly. Reading lists for various purposes are available at most libraries. Even if not followed closely, they can form a good framework for a winter's effort. A "Selected Reading" list for young engineers has been published by the Engineers' Council for Professional Development.

To get the most from your books it is invaluable to keep some sort of simple journal covering your reading. For each book an informal review of a few lines or a few paragraphs makes it easy for you to keep track of sources. It gives emphasis to important ideas and helps you to form more logical conclusions.

* See Chapter 19 for more on appraisals.

Most engineers take additional courses from local colleges or from their own companies in both technology and human development. A good rule, unless travel difficulties prevent, is to take a minimum of one one-semester course per year for at least the first ten years after college. Not all these courses need be difficult technical subjects. Cultural courses that husband and wife can take together are mutually beneficial. Many engineers do a great deal more than this and earn master's degrees in three or four years of after-hours classes.

Unfortunately, quite a few engineers recognize the need for further education but expend their energy by taking course after course in the plant or elsewhere with little coordination or advantage to themselves. It is difficult to criticize any educational effort. Getting a broad background is more often neglected than overdone. But courses for courses' sake can quickly become more of a salve to conscience than a help to your career. Particularly in heavily technical areas there is little point to taking anything deeper than a survey course unless it contributes to an advanced degree or to your specialty. For example, to take a first course in design with transistors could be a good exercise for almost any electrical engineer and many others. Presumably a good first course could open up a new field and a new way of thinking to the man educated solely in vacuum tube electronics. He could begin to appreciate the advantages and limitations of solid state devices. To take additional courses, however, would be a poor investment of time unless he were going to use the material. The project engineer will want his career efforts to add up to some significant result.

Company courses, particularly in the larger engineering firms, are offered in many useful areas. For example, a new engineering employee with advanced degrees found himself taking from his company the following in-plant courses during his first four years: professional engineering refresher, creative thinking, professional business management (two courses), effective presentation, financial statements, conference leadership. During the same period he taught effective presentation and professional business management and advanced two steps into his company's management.

Although company courses are usually more readily tied to immediate work benefits, select them carefully.

Professional Societies

It may be a very reasonable objection that the only people one meets at professional societies are engineers. Nevertheless professional society affiliation and active participation are major elements in the careers of many successful engineers.

Traditionally engineers have always shared their information and discoveries with each other. The professional society and its journal are a natural means for doing this.

New processes and devices developed by any one company may give it a competitive advantage in some industry. The engineer, of course, thoroughly respects the confidential nature of his relation with his employer or client in this case. But with a few exceptions these secrets are soon protected legally by patent applications and then possibly licensed to others, or more often their competitive advantage is gained by a few months of lead time. Then the information is shared with colleagues in a professional society. This may be at a local meeting, or sometimes at a regional or national meeting and by means of a journal paper.

All the members of a profession are concerned with and responsible for the profession as a whole. They exercise this responsibility through these professional societies, of which there are more than a hundred of interest to engineers.

The scope of American engineering societies is as varied as the specialized interests in the profession. Some are large with broad interests in a wide field—for example, the American Society of Civil Engineers (founded in 1852), The American Institute of Mining, Metallurgical and Petroleum Engineers (1871), The American Society of Mechanical Engineers (1880), The Institute of Electrical and Electronic Engineers (1884), The American Society for Engineering Education (1893), and The American Institute of Chemical Engineers (1908), all of which are more than 50 years old. Others, such as the Air Pollution Control Association, The Illuminating Engineering Society, and The National Association of Corrosion Engineers, are smaller and are devoted to relatively narrow fields.

Most engineers are members of the large society which covers their main area of interest. Many belong in addition to a small society in their own specialty or of one or more of the so-called "professional groups" through which large societies are beginning to break some of their own work down into specialties. Many professional societies work together in substantially overlapping areas. They form, dissolve, and combine as technical times change.*

* One of the best listings of North American technical societies is published by the Engineers' Joint Council (EJC), New York, under the title "Directory of Engineering Societies and Related Organizations," and is available at many libraries. This listing indicates the organization, purpose, and scope of specific technical groups.

You are concerned with your profession's present usefulness, efficiency, and service to the nation and humanity. You are also responsible for its growth and for providing continuation in the hands of the next generation. Young engineers are especially welcomed and sought for committee and other work. In addition to *technical* broadening and updating by association with other workers in a technical specialty, the engineer gains much *personally* by working with others, especially with older, more experienced colleagues. He can improve his oral and written presentation by exercising it in broader fields. Intelligent participation, even as a listener and voter, in society business affairs can begin to open up an understanding of the way his profession fits into the world around him.

The Status of Engineers

The concept of professionalism carries with it an idea of *status*. The status of a group is literally (and in a somewhat old-fashioned sense) its relative rank in society. From a more modern point of view, status is comprised of (a) what opinions other members of society hold about the group, and (b) what the group thinks of itself.

Historically status has been won by a group because of its importance and service to the rest of society. Status is usually manifested to some extent by privilege. Sometimes, but not always, it is shown by financial reward. When a group which has status ceases to serve, the status situation may continue unchanged for a surprisingly long time, but sooner or later it deteriorates.

Today many engineers and engineering groups are concerned with the status of our profession generally. Does it have its proper place in society, considering how important engineering really is and its potential for service? Do engineers have the rewards and esteem, the perquisites and privileges, that they should?

The profession is often compared, favorably or unfavorably, to law and medicine. More recently there has been concern that much credit due to engineering has gone to science. One of the reasons commonly advanced to explain (what seems to some) engineering's relatively inferior status is that the engineer, dealing in technical abstractions, does not come into effective contact with the public, as do the doctor and the lawyer. Instead the engineer deals with corporate or governmental employers or clients.

Most of the solutions proposed for this status problem would assume that the difficulty is somehow in the public mind. It is to be corrected through publicity, legislation, and other means external to the profession itself. Perhaps some effort directed at the public might help a little, particularly if it is aimed at *helping* people. They *should* reasonably understand an important element of their civilization. This collective work must be carried on genuinely as a service function rather than a self-serving activity.

Legislation which will genuinely protect the public rather than simply establish monopolistic groups will also be appropriate. All states now have engineering registration laws which provide for the legal recognition of properly trained and experienced engineers. The purpose is to ensure that only competent engineers can perform certain kinds of technical work, particularly work bearing directly on public safety. Since these laws exist and over 200,000 engineers are now registered under them, every engineer will want to take the steps required for registration.

If we are to support our profession and have a part in its collective voice it is important for us to work together. Otherwise governmental agencies will chart the course. The vehicle of licensing may play a progressively larger part in the development of engineering. There are still many faults in the system.

The only *major advance* in professional status that engineers can achieve, however, will come through (a) understanding the nature of engineering, and (b) practicing engineering on a completely professional basis. Then there will be no doubt *in their own minds* about the status of this vital work. Remember, engineering is *better providing for human need* through the intelligent application of technology. It is not technology for technology's sake. Professional practice must be creative, ethical, cooperative, self-motivated.

Engineers practicing on this basis will be giving, contributing to others, rather than seeking for themselves. In addition to doing technical project work they will inevitably enter more into their businesses and communities. They will contribute the kind of logical, cause-and-effect thinking, knowledgeable in physical things, that is more and more needed as engineering changes our civilization. They will associate in these endeavors with all kinds of their fellow citizens. These citizens will then hold an image of the engineering profession which corresponds with the one held by most of its practitioners, even as they probably do today.

Good Engineering Practices in Self-Development and Professionalism

1. The engineer looks on his career as a project requiring careful and deliberate guidance from himself as the project engineer.

2. The engineer recognizes that he must practice engineering as a profession, involving service to others, service to the profession itself, and ethical standards of conduct.
3. The engineer enthusiastically supports the interests and reputation of whatever employer or client he associates himself with.
4. When the engineer can no longer agree sufficiently with his employer or client to support him wholeheartedly, he breaks off the connection.
5. The engineer recognizes his special need for broad self-development as opposed to further technological training alone.
6. The engineer reads extensively and systematically throughout his career.
7. The engineer actively supports his professional society.

Poor Engineering Practices in Self-Development and Professionalism

1. The engineer considers his preparation complete when he graduates.
2. The engineer feels that his company will take care of him and guide him into the best path.
3. The engineer does nothing about career planning until cumulative dissatisfactions with a mediocre job finally drive him to look elsewhere.
4. The engineer looks on his work as mere wage earning, with no responsibility for dedication or service.
5. The engineer refuses most opportunities to participate in community affairs on grounds that other professionals are better prepared for this kind of service.
6. The engineer fails to recognize that excellent performance on his present job may be grossly inadequate preparation for the next promotion.
7. The engineer takes company and outside courses as a matter of habit with little regard for what they are adding up to.
8. The engineer takes no steps for earliest possible legal registration.

Chapter Eighteen

Creativity

Engineering creativity is an ability to conceive of a new combination of old things that can contribute to solving an engineering problem. It is important to note that you start with something already there.

It is a common misconception that inventions or other creative improvements are made more or less out of thin air. What a relief it is to recognize that you are not being called on to do the impossible or to deal in some kind of magic!

Imhotep, designer and builder of the first great pyramid at Saqqara, was one of the first identified engineers in history. "It was said that his plans 'descended to him from heaven, to the north of Memphis.' His counsel was 'as if he had inquired at the oracle of God.' "*

But innovation does not *start* with something new, even in Imhotep's case. It begins with the already known (and often quite familiar, as we will see). Then it proceeds to some change or improvement or new combination of the known. Often invention consists of combining a relatively new (but known) idea with an old idea or in an old situation.

Everybody needs creativity. Much relatively noncreative work can be handled excellently, too—like routine engineering technicians' work or detail drafting. But even here at least some element of creativity is useful and often encountered. At a little higher rung on the engineering ladder, advanced technicians and drafting designers can't get along without a considerable amount.

But if the professional engineer can't work creatively he is striking out. We saw in Chapter 1: if it isn't creative it isn't engineering. Therefore we are going to want to think about creativity in a little more detail. That is the topic of this chapter. But first let's be sure that nobody is going to neglect or strike out on his routine.

Every job has a considerable amount of routine which has to be handled carefully if the work is to be a success. Excellent performance in the more creative parts of a job coupled with poor performance on

---

* Richard S. Kirby et al., *Engineering in History*, McGraw-Hill, New York, 1956, p. 32.

the routine adds up to a poor job. The real art of professional performance, however, is to minimize the routine part. Provide for its effective handling—perhaps to a considerable extent by delegation or with a computer. Then put the extra time and effort available on the creative aspects of your work. You can even handle this "disposition of the routine" imaginatively and creatively.

You have seen that creativity doesn't start with something new. Another quite limiting misconception about engineers' work is that innovation always involves either (a) large and important projects, or (b) technical design ideas. Although a great deal of engineering creativity should naturally be concerned with technological things, it can't be limited to them. The creative engineer will express his imagination in many things from the design or conception of engineering devices (invention) to smoothing office procedures and effectively handling customers.

Creative versus Eccentric

Creative individuals and merely eccentric individuals are confused by some. What is the difference? Our definition of engineering creativity at the beginning of the chapter can help us here. You are looking into yourself for creativity which will contribute to the solution of engineering problems. By the definition, anything which hinders the work of an engineer cannot fit into that category.

You will find that anyone who is strongly creative is quite individualistic; we will see shortly why this is so. But eccentricity is individuality carried to the point where it interferes in some degree with the engineer's personal relations and therefore with his work.

We observed in Chapter 4 and again in Chapter 10 that a wise project engineer arranges to put most of his team's efforts on the critical factors. To do this he economizes elsewhere. We saw that standardization in design is a tremendously useful and economizing procedure, which as a rule is applied in the nonstrategic areas of the project.

The engineer applies these ideas to his creativity by innovating where it is especially needed and economizing in less important areas with more or less standard procedure. Now effective communication and over-all relations with others rest on common understanding. We would be foolish (like the eccentric) to innovate so widely in minor matters that we would be misunderstood or even mistrusted by others.

For most of us the creative effort, though far easier than many suppose it to be, requires a great deal of energy. Apply your effort where it is most needed.

Can You Learn Creativity?

A good deal of argument is heard in engineering circles about whether creativity can really be taught or developed. Isn't it an inherent trait that is or is not present in any given engineer? Perhaps it would be more accurate to ask, "Is the degree of creativity in any engineer fixed by factors no longer adjustable?"

As you might suspect, psychologists have given considerable attention to this important quality. But there seems to be no agreement as yet on a simple explanation of what creativity is.

In the meantime the engineer, as in many other aspects of his practice for the past several thousand years, must go forward. He has to find the best possible solutions to his problems today. He can seldom afford to wait for scientific theories to be developed, wonderfully useful as they are to him when they come along.

A good working hypothesis about creativity is that it is inherent in the individual by the time he enters our profession. It is widely distributed to most individuals and given in exceptional measure to a few. Even if this is the case, experience shows that for most of us the amount of our inherent creativity that we *use* can be greatly increased by effort and practice. In fact lack of imagination and fear of invention appear to be mostly *habits of thought.*

I have had the good fortune to participate in creative engineering courses in industry and to work closely with the graduates of creative engineering training programs. There is no question at all that the majority of people trained in these courses *expressed* creativity to a significantly higher degree as a result of their schooling.

Most engineers in these programs (a) discover that they can work creatively to a high degree, (b) gain some experience in creatively solving problems, and (c) learn tools of creative and imaginative thinking which they can apply successfully to everyday problems after graduation. As in other training, creativity seems to decline as time passes unless it is actively and continually practiced.

Your Choice

With these facts before him, every engineer has a clear choice between a mediocre career (in which he does not consider himself to be capable of much creativity) and the satisfaction of a far more successful career (in which innovation is a major element). Some helpful books are available on the subject. The pursuit and development of creativity must be a life-long interest. But let's look briefly at the fundamentals.

Some Fundamentals

If creativity is an inherent quality and the problem is simply to express it, techniques for improvement will primarily involve *tearing down or reducing the inhibitions which stand in the way.* We will consider these important inhibitors: lack of information, habit, pride of knowledge, fixations, experts, education.

Have You Enough Facts?

Since the practice of engineering is based on technology, you can do nothing without extensive, appropriate technical information on the problems with which you are dealing. We have seen that creativity is the combining of old ideas or known things into new patterns. Without an appropriate stock of the already known you are in no position to be creative in solving a new problem.

Experienced engineers find that, when they are bogged down after considerable effort, the reason is almost always lack of sufficient information. In that situation stop working for a while and go learn more facts.

Use Habit Wisely

Habits are servants as well as masters. It is often remarked how much of our daily activities are handled by habit instead of thought. From clothes selection and dressing procedures in the morning through driving and parking, greeting associates, food selection and eating, and on to details of daily work, we move by habit. This is part of our personal economical standardization of noncritical activity. If there is no need to think seriously on such subjects, why make the effort?

Of course, some of these items should be looked at more carefully, at least once in a while. We all have stodgy associates whose manners and dress are becoming a liability to them and whose absent-mindedness will soon be inefficiency.

How many stodgy engineering habits of your own do you fail to consider? Do you habitually look at most of your problems in the same old way? Are your analysis techniques never varied? How many brand new engineering procedures have you made your own in the last three years? How many radically new devices have been incorporated into your designs?

Habit is helpful at times but can be dangerous. If you consistently follow habit, you can hardly be creative.

Cultivate Correct Attitudes

Creativity requires both confidence and humility. These qualities have no element of mutual inconsistency. Humility allows you to recognize that you depend on laws and concepts in your engineering work (both technical and human) that belong to everyone. You can know only a fraction of all the things there are to know. Help and ideas can come to you from anyone and from most unusual sources.

Thus equipped, you are in a position to feel confidence in your ability to search out the ideas you need. Also, you have confidence in the impersonal laws of technology and in the engineering procedures available to you.

See Through Your Fixations

Fixations are blocks to new ideas caused by strong mental associations between objects or thoughts. An example cited by E. K. Von Fange[*] illustrates this definition. Subjects in a test were asked to devise a way of mounting some candles on a door with materials provided. In addition to the candles one group was given several small open boxes and a few tacks. These subjects all quickly conceived of using the boxes as platforms for the candles. Another group was provided with the same materials but with the tacks and candles in the boxes. Under this condition many of the subjects were unable to conceive of using the boxes as mounting platforms for the candles. It apparently was clear to them that the purpose of the boxes was simply to hold the candles and tacks. This fixation prevented their conceiving of using the boxes for anything else.

Faced with a design or development (or any other) problem, look for fixations. Put them aside before you begin work. "How else can I use this or do this or make this or relate this?" you ask yourself.

Beware of Experts

Experts and expertise are major inhibitors to creativity. The expert may be a person who is skilled in the art that you think is involved in your problem area. Or the expert may be a textbook or magazine article. Even more subtly the "expert" may be some preconceived and "sacred" concept about the problem in your own thinking. Whatever

[*] Eugene K. Von Fange, *Professional Creativity*, Prentice-Hall, Englewood Cliffs, N.J., 1959, p. 31.

it is, the tendency for an engineer with a problem is to (a) give up on it himself in favor of the expert, and (b) accept whatever the expert says without further thought, even a verdict that the problem is insoluble.

This is not to say, however, that specialists and experts do not have a use. With deeper and more up-to-date knowledge and experience in the problem area a specialist should be in a good position to be creative, having more to start from. It is foolish for an engineer not to avail himself of specialized knowledge when he can. However, experts are frequently blinded by their own experience and detailed knowledge. For this reason capable engineers often develop a strong skepticism about experts or even any specialist.

A vice president of a large company kept score in his early career on the correctness of his colleagues who spoke authoritatively on any subject. In the instances where he was able to judge he found them only 60% correct. Thus it would appear that two experts would agree with each other perhaps 36% of the time.*

Although these judgments may be harsh, they certainly illustrate the need for the creative engineer to guard his thinking and project from the experts he needs to consult. Experts in many instances consider themselves to be the last word on the question. Presumably no further ideation on anyone's part need take place. Of course, the most useful specialists do not take this omniscient attitude.

Education Is Dangerous

The generally inhibiting effect of education has been remarked on by many, especially by people who themselves have been extremely creative. Albert Einstein felt that he had to make an effort to forget most of what was taught him so that it would not stifle his thinking. The following quotation from Charles Kettering, a prominent development engineer in the automotive field and vice president of General Motors, is typical:

"Some years ago a survey was made in which it was shown that, if a person had an engineering or scientific education, the probability of his making an invention was only about half as great as if he did not have that specialized training. . . . An inventor is a fellow who doesn't take his education too seriously."†

* *Ibid.*, p. 64.
† From an address to the American Society of Mechanical Engineers in 1943, quoted in E. K. Von Fange, *op. cit.*, p. 224.

Thus the engineer is in a dilemma. If he thoroughly educates himself technologically, according to many very successful and intelligent innovators he is stifling his creativity. But we know also that creativity is built on knowledge. If he does not thoroughly educate himself technologically, he cannot hope to go very far in engineering innovation. The solution lies in recognizing that it is not real knowledge which stifles, but erroneous attitudes toward the small accumulation of knowledge that any one engineer can have. As Kettering indicates, the effective engineer does not take his past education too seriously.

Breaking Through

What should the engineer who recognizes these inhibitors to his creativity do about them? Probably the recognition itself and consequent alertness in his day-to-day work are his best protection.

However, various more or less artificial procedures have been worked out to help. Some are common-sense methods, while others are controversial. Some work with almost everyone. Others are of no help at all to many people.

Creativity is an individual and personal thing. Each must by trial work out his own methods for improving his work. Certainly the first requirement is a continual effort to find new and different solutions to all kinds of problems.

Incubation

The actual process of ideation seems to be widely different in different people or even in the same person at different times. In most cases it is essentially unconscious. The need or problem is put mentally beside the facts, the known facts with which creativity begins. Then something happens. Solution ideas begin to come.

Most observers agree that hard and long effort up to the point of mental exhaustion is a major factor. Often this effort does not yield results immediately, but it tends to blast loose mental habits and preconceptions. Then the engineer's thinking is freed to find a solution. It is well known that some people sleep after this effort and find a solution coming to them the next day. Others wake in the night with the answer to the problem.

The generalization and extension of engineering problem solutions discussed in Chapter 6 is so often fruitful because it follows a long, hard mental effort. The engineer's thinking has been forced onto new and less circumscribed ground.

Check Lists

The use of check lists is another way of forcing thought beyond habitual modes. Some people use catalogs or the yellow pages of the telephone directory. Alex Osborn describes a number of lists or question techniques.* Many engineers devise their own lists.

Brainstorming

Another well-known technique is brainstorming. Properly used, it can be invaluable, but quite a few people find themselves unable to participate in it. It is based on two ideas: (a) criticism and the instinctive fear of criticism are strong inhibitors of ideas; (b) one new and unusual idea often loosens up thinking so that immediately another is born.

In the usual form the group which is to brainstorm sits in front of a blackboard. One of the group is appointed to list ideas on the board. Until the group becomes quite expert it is best that the chalk man not contribute ideas. No criticism or comment of any kind is allowed on ideas. Group members simply call ideas for solution of the problem assigned, to the chalk man, who writes every one down as briefly as possible. In a good session the ideas come so fast that the chalk man is usually behind.

At first, the wilder the suggestions the better, in order to loosen up group thinking. There is no need for the ideas to be practical or even technically feasible as stated. Surprisingly, wild and impractical ideas will sometimes beget further ideas that remove the impracticality. The inhibitions in one person's thinking are not necessarily shared by others. Ideas old to one participant are new to others and evoke a response.

Typically a hundred ideas so generated in the space of ten to fifteen minutes may contain two or three worth further work and consideration. Thus this activity does not solve problems; it only provides new ideas, sometimes radical new ideas, for further investigation. In many cases it also serves to list comprehensively the various approaches possible to an engineering problem. *After* the brainstorming session the group provides for some critical evaluation of its list of ideas—an evaluation which may set a direction for another, similar ideation session.

* Alex Osborn, *Applied Imagination,* Charles Scribner's Sons, New York, 1953, p. 271 seq.

For those who are mentally flexible enough to use it, this technique can be quite valuable. One-man sessions with a typewriter can give surprising results, though lacking the stimulus of other people's ideas.

This technique has some disadvantages for engineers. Hopefully, most members of our profession are not given to voicing opinions or ideas lightly. When an engineer comments professionally, it should be assumable that he speaks from knowledge. It is justifiably hard to turn this attitude off for a brainstorming session. Certainly the brainstorming concept must not be carried into practice.

A further group disadvantage occurs in patent matters, where it becomes difficult to establish proper credit for ideas. However, the original group, if accustomed to working with each other (and if commonly motivated), can usually solve this problem.

A consistent and confident effort to look creatively at all problems is hard to beat. In your regular self-appraisals you will want to take a thorough look at this characteristic and to make realistic plans to increase your own expression of it. You will not be very creative unless you systematically and deliberately practice creativity.

Good Engineering Practices in Connection with Creativity

1. The engineer recognizes that he will not be very creative in his work unless he goes about it deliberately and systematically.
2. The engineer works creatively in everything he does, including administrative matters, instead of limiting his ideation to technical things.
3. The engineer minimizes the nonprofessional portion of his work by providing carefully for routine to be taken care of properly, possibly by delegation.
4. The engineer uses habit as a servant, but also recognizes its danger as an inhibitor of ideas.
5. In using experts and consultants, the engineer avoids letting them do his thinking for him or make his decisions.
6. The engineer takes advantage of opportunity for a training course in creative methods.

Poor Engineering Practices in Connection with Creativity

1. The engineer is satisfied in his work to go along unprofessionally, doing things the way he has always done them.
2. The engineer feels that creativity is a matter of endowment alone.

3. The engineer fails to express significant creativity in some area because his technological understanding there is too meager.
4. In his attempt to express creativity the engineer neglects the routine portions of his job.
5. The engineer assumes that creativity involves "something out of nothing" instead of recognizing that it must start with known ideas.
6. Failing to recognize that creativity is an intensely individual thing, the engineer assumes that the successful procedures of others will always get the same results for him.

Chapter Nineteen

The Engineering Manager

Whether you work for a manager as his employee or are retained as his consultant you will need to understand the manager's function and some of his principal problems to support him effectively and professionally. This chapter is intended, not as a treatise on engineering management, but rather as an orientation in the manager's function *for those who will be managed.*

Management, of course, is primarily a matter of getting a job done through the efforts of others. We saw in Chapter 14 the relation between a project engineer and his unit manager. Figure 14-1 relates both these engineers to the engineering enterprise as a whole. Of course the unit manager is ultimately responsible for all the work in his unit. He normally delegates authority for the day-to-day direction of projects and puts most of his own effort into longer-range work for his entire unit. He is directly responsible to his own manager for all the project work, however, and must guide and support his project engineers as required.

Project Engineer Works in the Present

A useful way to differentiate between the work of project engineers and unit managers is in terms of the time span of their principal responsibilities. The project engineer's effort can hardly afford to stray very far beyond his project. Of course he has all the normal concerns of any engineer and will take a strong interest in the progress of his unit and the company as a whole. But essentially all of his effort and time will be required on the project.

When a project engineer begins to put large amounts of his time on things in the future—unless that is what the project is concerned with—he is probably getting into trouble because his present responsibility will suffer.

Manager Works in The Future

The unit manager, on the other hand, is primarily interested in how his projects and their results add together to promote the company's future interest in his technical area. The unit may be responsible, for example, for designing all the radar equipment that his company sells. He will be looking ahead to see what the future developments in radar technology may be. He tries to see how the company can effectively contribute to them.

He will be looking for the right opportunity for his unit's next design projects. Also, he will want to be sure that he has (or is developing) the right combination of technical talents and specialties in his unit to be ready for the next job.

He helps his project engineers to whatever degree they need assistance, but if he is spending a substantial part of his time doing detail work on projects, he is probably getting into trouble. The unit is not being prepared for its future projects.

Supervisors

Sometimes another level of supervision is inserted between the first-line manager and his men. Persons in this intermediate level are often called supervisors. You will particularly encounter supervisors in production organization, large bookkeeping operations, and the like, where there are many nonprofessional employees to be directed effectively.

Although the term "supervisor" is still encountered in some engineering design groups, it has come to mean first-level manager in most cases. The idea of literally "supervising" the detail work of an engineer is not a professional concept. An engineer who needs this kind of supervision had better turn in his calculator.

Since new engineers usually work as individual contributors on project work, their interests are properly concentrated on today's project. They have to make an effort to adjust their thinking beyond the present in considering what the manager is doing. It is worth learning to do this early in your career, however, for a considerable amount of valuable vicarious experience is available to anyone who watches his manager's functioning carefully and thoughtfully.

The Manager's Daily Work

The work of a manager has been summarized effectively in four terms: plan, organize, integrate, and measure.° Although there are other ways

___
° This terminology is used in the General Electric Company.

to express the same idea, I can find none that is at once as clear and as concise.*

Planning

Planning is deciding what is to be done. The manager plans a program of activity for his unit against time. He should probably know quite well what his group will be doing in three months. Plans for a year from now are more indefinite. Plans for two to five years in the future must be very general and tentative. Beyond five years the plan is probably a mental estimate developed as the manager reads technical journals and talks to others.

Just as the project engineer knows that his schedules will be changed as time passes and unforeseen developments occur, so the manager is ready to alter his plans as required. They are continually under review and revision. But plan he must if his unit is to be prepared for technological and market changes.

Perhaps a small merchant might expect to continue for a long time at his present location, using the same sales force, buying from the same suppliers, stocking about the same line of merchandise, selling to the same customers. He can get along with substantially no planning. (or is the center of the city decaying for lack of parking areas and his business migrating to his competitors' new stores in suburban shopping centers?) But in any business concerned with technology rapid change forces planning.

The major problem in effective planning is to provide for a series of activities which is (a) contributing to the goals of the entire enterprise, (b) self-supporting, and (c) reasonably continuous. A secondary problem is to determine the quantity and type of people and facilities required to carry out these planned activities.

Organizing

Organizing is breaking up the planned activities into man-sized jobs and deciding who will do them. The principle here is to organize from the top down. First, project engineers are selected, assigned, and then used to help finish the detailed job planning. Flexibility is an important requirement of all organizing since it is often not known exactly which persons will be available or how the job requirements will change.

---

* Readers working as individual contributors will find it interesting to also apply these four ideas, as they are briefly developed here, to managing the more or less separate parts of their own work or projects. How much modification do you need to apply them for this purpose?

Engineers with narrow interests hired into the unit for an anticipated job may constitute an expensive liability if the job does not come in. Engineers who can work in a number of areas, on the other hand, are a real asset to a manager. The smaller his unit, the more of his engineers will have to possess this kind of flexibility.

Engineer-Manager Conflicts

Engineers and their managers normally get along quite well. Much of what trouble there is between them is caused by organizational problems. The engineer is working himself into an interesting specialty; he learned a lot about it on his last job and wants to continue with it. But the manager needs him in another kind of work. The engineer sees his career plans and interests being blighted by neglect. The manager feels that he cannot afford to support the engineer on research or proposal work in the old specialty and also needs to get his new project under way.

The solution to this kind of problem lies in both individuals together taking an honest look at unit plans. If reasonably good planning has been done, they can see what the prospects for a near-future return to the old specialty are. If there are no prospects, the individual engineer has the choice of developing some other specialty or taking his talents elsewhere. If the plans appear to favor additional work later in the desired specialty, the engineer can take his chances, regard the new work as a broadening experience, and continue on the side with his first interest, with the encouragement of his manager. But support his manager he must as long as he chooses to stay with him.

Of course, at a time like this, even the best of managers are biased in favor of the jobs they must get done. A professional engineer runs his own career. He will have to make his own decisions. If for financial reasons continuity of employment is essential, there may be some very real constraints on these decisions. (It is most desirable for professionals to work themselves rapidly into a position where these constraints are minimal.)

A troubling experience that most managers have at one time or another with engineers grows out of the situation just discussed. The manager needs the engineer on the new job. The engineer doesn't want to do that kind of work and finally decides that he won't accept the job. Then he takes the attitude that it is the manager's responsibility to find him another place either in the company or outside of it. Most managers are willing to help where they can in these matters, but it is unrealistic to expect them to take responsibility for their engineers' careers or to make major decisions for them.

Integrating

The integrating function of a manager is his daily activity of smoothing out the working of his organizations and people. The best plans change rapidly. Their details are never adequate to cover all the contingencies in a project situation. The manager must help his project engineers and others over rough spots. He approves at least some of the major decisions. He supports project engineers in areas beyond their authority.

His engineers have most of their contact with the boss in this phase of his activities. A good manager looks over his operations regularly, including, usually, physical inspections.

Much management theory and practice fail because they envision a more or less stable situation in which all members of a unit know their jobs and have been doing them for several years. In a real-life engineering organization changes come too fast for much stability to set in. Hence the manager has a continual training task to perform. Most training is informal and comes about through the manager's helping an engineer to correct mistakes that he has made or is about to make. Or the manager steps into a difficult situation and by example shows how to handle it.

Although integration is an important and even indispensable part of any manager's work, ideally he will minimize it by excellent performance in his other three functions. The usual mark of poor management is spending the majority of time in detailed integrating work to the neglect of planning and measuring. Much poor management work is the kind of activity referred to as "fire-fighting," that is, taking care of serious troubles as they arise. A manager who spends most of his time doing this must neglect his planning and measuring and thereby is sowing future troubles that will require him to be an even busier fire fighter.

Measuring

Measuring is the fourth management function on the list but is really a continuous process. In fact all four of the functions apply to each other and are recycled continuously in complete analogy to the five engineering functions described in Chapter 1.

As the project engineer carefully examines the results of his various efforts to see what should be augmented or what curtailed or eliminated, and how the direction of his team's efforts should be changed, so the manager continually measures or appraises the progress and results of all his plans, organizations, projects, and people.

Which plans are not working out as desired? Can they be adjusted to accomplish the end sought, or should they be abandoned? Is a particular organization operating efficiently? Can it be improved? Is there some-thing to be learned from this result that could be applied to other orga-nizations? Where are the project breakthroughs occurring? What long-range technical plans should be revised as a consequence?

Performance Appraisals

Measurement of his people helps the manager to establish needed and realistic training programs. He plans for hiring new kinds of talent when needed. He plans to eliminate those who are not going to be able to pull their full weight or are no longer needed. He selects those who can be considered for promotion to bigger responsibilities. He ad-justs salaries.

Since his people are professionals, the engineering manager can use this measurement to guide and stimulate them directly. Most large com-panies have a policy of giving to each professional employee a written appraisal of his work at regular intervals, perhaps once a year. Often the appraisals are quite detailed and include recommendations for self development to overcome weaknesses or build on strengths.

An engineer should take his numerical ratings with a grain of salt. The only really significant aspects about quantitative ratings are (a) the *relative* markings on the various qualities or performance factors rated, and (b) the overall rating *compared to other engineers* in the unit. This last point is usually impossible to determine and will have to be estimated from the amount of salary increase given in the period after the appraisal.

Suppose that all your ratings except two are in the top bracket of the appraisal form, but these two are next to the top in a column marked "excellent." It is these two performance factors, then, that the engineer is interested in. Can he improve them? Why does his boss feel that he is relatively deficient in them?

Instead, perhaps most ratings are in the next-to-the-top bracket but two are at the top. The engineer is interested in these two. What can he do to exploit them further? What is there about his boss's work that makes them valuable?

In other words, the engineer is primarily interested in finding the two or three qualities which the appraisal indicates are high and the two or three which are relatively low. He attempts to build on his strengths and improve or circumvent his weaknesses.

In some cases there is an "appraisal conference" along with the written evaluation. Many engineers and managers dread this discussion. (Some psychologists, in fact, feel that such a conference is destructive rather than helpful.) These talks can be very helpful to both the man and his manager, however, if each has the right attitude toward them.

It is important for each to recognize that the appraisal is *one man's very imperfect picture* of what another man's performance has been for him over the past year. It is *not* an appraisal of the man himself. It is merely a rating of what his manager *thinks* that the man has done for him.

Professionals do not ethically tear down each other's character, qualities, or capabilities, but managers must decide whether raises will be given, employment continued and accepted, or promotions made. And the more experienced always go out of their way to help the newcomers.

We saw in Chapter 15 that an effective engineer must get to know his manager in order to support him intelligently. The appraisal conference is a good, if stressful, time to learn about him. Ask him questions about the ratings. Try to get past the formalities and words to find out what he really thinks and recommends. Start him talking on these subjects.

A good manager will help his engineers to be at ease and to get the most from such a conference, but in many cases the situation is harder for the manager than for the man he is appraising. Many managers recognize, after a sincere and honest effort at appraisal, that they really have little knowledge of the man and his performance to go on. During the conference they feel on the defensive about this inadequacy. When an engineer shows by his questions that he is (a) aware of the nature and difficulty of the appraisal process, and (b) professionally interested in learning all he can to improve himself, this puts the manager at his ease and allows him to become more helpful.

One of the most difficult things for a professional to obtain is an honest opinion (right or wrong) about his work from someone else. He should not miss the chance. He will probably not agree with all his manager's opinions, but rather than argue the points these two professionals simply examine their different viewpoints. It is not a question of man and master (as it nearly is with nonprofessionals) but of a professional man and the professional manager for whom he is presently working.

Budgets

A major source of management problems is symbolized by the word "budget." At regular intervals, usually once a year, the manager must submit to his own manager a financial plan for his next year's operations.

We saw in connection with project control that a project budget or money schedule is essential. Similarly with an engineering unit it is inconceivable that intelligent planning and organizing can be done without a budget. But we have seen the chance nature of much engineering project work. At the unit level this probabilistic situation deteriorates further. There is doubt as to which outstanding proposals will be awarded, how many new engineers can be hired next summer, what the company policy toward summer employees will be this year, what developments the competitors will come up with, and so on. Thus engineering budgets seem particularly difficult to make and follow realistically.

Attitudes toward budgets vary from one company to another and within the same company from time to time. In some organizations it seems almost more important to keep precisely on budget than to get the project work done. Others appear to make a budget as an approximate planning exercise and then forget it for the rest of the year.

A lot of the variation in budget attitudes depends on financial health. If an organization is doing well, accomplishing useful and therefore profitable engineering, supporting itself easily, and looking toward a promising future, budgets seem to have relatively little importance. But when a group is in financial difficulties—and most are at times—budgets naturally assume a larger role.

It is hard sometimes for engineers to understand why a penny-pinching approach is suddenly imposed. These crises may have their origins in poor management farther up the line. They may be due to a national business decline or some new competitive situation. But a financial crisis is a crisis regardless of its cause.

Businesses can no more operate unscathed in the red than individuals can. To a considerable degree red ink is red ink regardless of where the decimal point is. When a unit manager is required by his own manager to meet certain drastic financial curtailments, his engineers have no choice but to support him loyally.

Foolish as some areas of savings seem to be technically—cutting off promising development projects, for example—they are judged somewhere in the company's management to be less foolish than not being able to pay salaries. Of course, if this sort of thing happens too often, the alert engineer may question the amount of useful work that he can accomplish in that organization in the future.

Although it is generally necessary to meet budget requirements, to take this as a principal goal is usually fatal to an organization. In addition to meeting the budget some useful engineering work must be done.

Managers' Rewards

The previous discussion of managers' trials and tribulations is intended to help you understand your boss a little better, but it shouldn't discourage you about managing. Managers have many satisfactions.

In the first place, they are paid quite well in most engineering businesses. They often tell each other that they are paid to take care of trouble and to solve problems. And if they delegate and organize efficiently, most of their integrating work will be just that—trouble shooting.

Bring your problems to them; that's what managers are for. Be sure of course, to come with your own best solution. Then in most cases the boss can bless it, and that difficulty is solved. If he doesn't like your idea, he can give some guidance and you have learned something. (Of course on project matters you go to the project engineer.)

One of the strongest satisfactions a manager has is working with his project engineers to develop them into better engineers and better leaders. The transition from individual contributor to total responsibility is a big one. The thoughtful manager will find much that he can do to help a new project engineer learn quickly, but he is careful not to interfere with him.

Developing the group as a whole and nurturing the specialities and other capabilities of the individual engineers constitute a challenging task. We have looked at the formal appraisal from the man's point of view, but to the manager this is just one step in his continuing effort to improve and develop his engineers and other contributors. As a professional dealing with others he feels a sense of obligation to help and to provide an atmosphere which encourages their development. But he cannot take (and is not interested in taking) any responsibility for their careers or career decisions. He is primarily interested in developing them as a part of his unit and as useful subordinate leaders for his own bigger assignments in the future.

If a man has a wide interest in technology, he will find satisfaction in having a manager's part in influencing major technical decisions and developments even though he can no longer afford the time to regularly explore technological depths. In his endless planning and search for a right direction for his unit's future he may consider a vast sweep of technology and also of potential applications to fill human needs.

Since he was an outstanding man in his days as an individual contributor, he is still possibly the best or one of the best engineers in his own unit. When time permits, his subordinates can learn much about engineering practice from him—for example, when they are on business trips with him.

As the years pass, he will get to know more about the total business in which his company is engaged and will help to make the tactical decisions that run it. Of course, if he is the president or general manager, he can have the satisfaction of running his own show.

Summary

1. An engineer must have a reasonable understanding of his manager's work if he is to support him intelligently and professionally.
2. Project engineers and their associates are primarily concerned with their immediate projects. Managers are normally involved with longer-range planning and the integration of a number of project results into the unit's overall goals.
3. Managers' work can be conveniently divided into four activities: planning, organizing, integrating, measuring.
4. Performance appraisals are simply one man's opinion of another's work. They are an excellent opportunity for the appraised engineer to obtain a reading on himself but are not necessarily to be taken as exact truth. The appraisal and the appraisal conference provide a good opportunity for an engineer to get to know his manager better.
5. Business organizations must live within money budgets, just as individuals must cope with income limitations. Many of the seemingly nonsensical things that managers do can be traced to this necessity.
6. There are many satisfactions in management work for the engineer who is capable of it.

INDEX